PRAISE FOR DINO

As a 21-year-old trying to build ...., simultaneously overrun with anxiety and overwhelmed with grief and fear about what little promise there is for a future on Earth. It's SO easy to let fear take over and think there's nothing much I can do: if this was REALLY as big as they say it is, surely someone would do something. But that's exactly the problem, the people we're supposed to look to for direction in times of need are utterly blinded by greed and money. For so long we've all been following what the media and news fed us, while staying in the dark about the realities of climate change and global warming. Meanwhile, the ones who hold the highest office of the United States and who were sworn to make the best choices on our behalf have ultimately betrayed us. I found it so interesting that my grandmother wove her life story around the story of climate change. It was powerful having the anecdotes in which she was buying a car or doing some other, rather ordinary task, while unbeknownst to her the big oil industries were working to cover up any word of climate change. Now she shows me how vital it is that those of us who understand the magnitude of the crisis take action so that, hopefully, we will be able to outweigh the actions of those impeding our progress."

–Channing Pooley, Author's Granddaughter

"*Dinosaur Dreaming* is a beautiful book, an exceptionally important one, and it comes just in the nick of time. Collins-Ranadive weaves her own personal story into her narrative about 'our climate moment.' Here, in compelling story and vivid image, complicated realities at last make sense and invite smart conclusions. This book is a must read right now."

–Barbara Child, author of *Memories of a Vietnam Veteran*

"Collins-Ranadive writes a memoir that sets an example of the moral accountability and right actions that climate change demands of people and societies of good will. Her concisely summarized research and clear language in explaining the science and politics of this complex issue are exemplary. Her call to readers is to stand strong, with open eyes, and dream and co-create a different future. *Dinosaur Dreaming* shows us how to confront the greatest challenge humanity has known with awareness, hope, guts and grace."

–Elise Houghton, Environmental/Climate Education Advocate

"Collins-Ranadive skillfully weaves together science, story, and soulful imagination to provide a vivid snapshot of our present predicaments and powerful lifelines in addressing the biggest challenge humanity has ever faced."

–Aly Tharp, Program Director,
Unitarian Universalist Ministry for Earth

"This book is a manifesto of action for all of us that care about the future of our planet. As a teacher, I am inspired and buoyed by the young people to address the realities of climate science. For them, we must all lead with our values aligned to ensure our Earth's health for all who come after us. Gail is articulate in weaving her authentic personal journey with the urgency of 'Telling the truth' of our climate crisis."

–Theo Small, Teacher and Union Leader

"What we once referred to as "climate change" has been so ignored that it has now become a "climate crisis." Now ignorance is no longer bliss. Slowly we've realized that preserving the earth is to preserve our lives, while destroying earth is to destroy ourselves. Now the question is, is it too little and too late?"

–Donald Winslow, Journalist and Painter

"In this luminous book, Collins-Ranadive teaches us how to see the climate crisis through the eyes of a wise elder. Her life's rich and inspiring stories, from motherhood to climate ministry, pull us into the discussion in a new way. She has a way of opening our hearts so we can begin to challenge ourselves to see the clear truth of the problem yet not be defeated. Her careful research educates us by taking us on a creative journey through a rainbow prism to understand what brought us to this place of planetary emergency. She and her partner Milt Hetrick have been brave stewards in their work and inspire us all to 'transform our sense of hopelessness into activism on behalf of all that we cherish.'"

–Paddy McClelland, Co-founder, Wall of Women

Glacier calving this finite blue marble

Eye to eye
with St. Francis
bird the sun shares

heavy impact
of the waterfall
no longer falling

warming up to
the idea of climate change
by degree

–Jackie Maugh Robinson, Haiku Poet

GAIL COLLINS-RANADIVE

DINOSAUR DREAMING

OUR CLIMATE MOMENT

HOMEBOUND PUBLICATIONS

Ensuring the mainstream isn't the only stream

Postal Box 1442, Pawcatuck, Connecticut 06379-1442
www.homeboundpublications.com

© 2020 · Text by Gail Collins-Ranadive

Quantity sales. Special discounts are available on quantity purchases by corporations, associations, bookstores and others. For details, contact the publisher or visit wholesalers such as Ingram or Baker & Taylor.

ISBN · 9781947003668
First Edition
Front Cover Image © Map by AnSuArt
Interior and Cover Designed by Leslie M. Browning

Printed in the United States of America
10 9 8 7 6 5 4 3 2 1

"The elephant in the room is the room itself, the biosphere in which all life currently known to exist in the universe is enclosed, and on which it all depends, the biosphere now devastated by climate change, with far more change to come. The scale is not like anything human beings have faced and journalists have reported on, except perhaps the threat of all-out nuclear war—and that was something that might happen, not something that is happening. Climate change is here, and it is changing everything. It is bigger than anything else, because it is everything, for the imaginable future."

—Rebecca Solnit

For

MacKenzie,

Channing,

Alex,

Addison,

and Liam,

of course.

CONTENTS

FOREWORD

In a recurring nightmare, I am standing in a linked-arm chain of climate-conscious citizens. We are lined up along the cliff-edge of extinction, facing off against the perpetrators of the greatest crime against humanity: the polluters and profiteers who have known for decades that their products would warm the planet, change the climate, and harm life on earth by overloading our atmosphere with CO_2 emissions.

Now they've completely hijacked our government that's supposed to be of, for, and by we the people and are poised to exploit every last ounce of fossilized sunlight for political power and private profit.

The only thing standing in their way is us...anyone and everyone who knows we are at a crossroads unlike any other in human history.

In my nightmare, I fear I'll be a weak link in the line of resistance, a place where the climate deniers and liars break through, with those they have duped following as mindless as lemmings headed straight for the cliff. As the juggernaut of existential death presses towards us, I worry that my grief over what's being lost will swallow me whole. I hang on to dear life despite throbbing arms. Surely this is just a bad dream!

Wait! Is that a snowball being lobbed in our direction, that ultimate mockery aimed at the scientific reality of anthropogenic climate disruption? Now we're being pelted with snowballs, each of them representing the latest assault on our attempts to deal with the climate crisis: the international climate agreement, the national clean power plan, the Keystone XL and Dakota Access pipelines, state and local renewable energy portfolios, the mileage standards for cars that would reduce emissions.

As the list grows to include offshore and arctic drilling, and leasing public land for fossil fuel exploration and exploitation, snowballs smack us and break apart at our feet, hissing like snakes. A growing pile of slush is turning into mud that is so slick and slippery we're losing our footing as the fossil fools surge forward. I struggle to keep my balance. Or wake up.

This can't really be happening! Somewhere in my nightmare my mind knows it is simply processing the latest findings that warn we have barely the rest on my lifetime to turn this planetary crisis around.

Submerging into the nightmare again, I catch the glimpse of a rainbow. As if the Sun has our backs, a prism of color streaks across the dark cloud above the heads of those who'd destroy our world. They cannot see it.

The rainbow carries me into my waking life, until, on my morning walk, an actual rainbow graces the misty sky and awakens my imagination. Rainbows have long symbolized hope for humankind, from the Greek goddess Iris carrying messages along its bridge between heaven and earth to the Hebrew god promising Noah to never destroy the earth again.

This time it's we humans who are destroying the world. With tipping points and feedback loops looming, we're at a place of no return to the "just right" climate that brought us forth and sustains us. And I've grown so weary and become so demoralized over this that my inner and outer worlds have become heavy, a hopeless grey. I'm ready to give in and give up.

Then all at once my grey spirit shape-shifts into a dove that whistles by. Is it in my dream or in my day? It doesn't matter. In the dove I see a dinosaur that learned how to fly, and this remnant of the last mass extinction begs to be a metaphor to help prevent the next one.

Options for responding include flight, fight, and freeze, plus the uniquely human ones of figure out and fix it. A lifelong student of nature and human nature, I choose to believe that our species can and will wake up and take action.

Never mind that the perpetrators would have us accept that it's too late to do anything that would disrupt their profits.

We created this crisis and we can solve it. My arms stiffen in defiance of all odds. Don't dare tell me we won't rise to this challenge! Fear turns into fury and awakens a fierceness that frightens me.

Focusing on the rainbow as it refracts seven facets of the problem, I see the solutions that are embedded within each. We don't have to go the way of the dinosaurs, and the human experiment need not end in the blink of this cosmic moment!

ONE. RED
WARNING

all years ct smell (handwritten)

"We're the first generation to feel the impact of climate change, and the last generation that can do something about it."

—Gov. Jay Inslee

HOW DID THE HUMAN SPECIES get itself into this existential mess? Sadly, my own generation has been complicit in creating the predicament.

I confess I was so completely oblivious to the problem that I can't even recall seeing the big green Brontosaurus on display at the New York World's Fair back in 1964. Did I even visit the oil company pavilion that featured their signature mascot and showcased nine life-size replicas of dinosaurs as a ploy for promoting their gasoline?

Frankly, dinosaurs probably wouldn't have grabbed my attention or remained in my memory; those huge beasts were something my brothers were into and comprised the biggest words in their vocabularies. As for my sisters and me, we simply wouldn't have been impressed with putting a dinosaur anywhere, let alone in a gas tank, a marketing claim based on the false premise that the world's oil reserves were laid down when dinosaurs roamed the earth.

1

Yet even back when I was a teen there was scientific evidence suggesting the dangers of burning gasoline, wherever it came from. In fact, research on the effect of CO_2 on the climate had begun as early as 1824, when the atmospheric greenhouse effect was first discovered. According to the information found in ice cores, for 800,000 years after the age of the dinosaurs atmospheric CO_2 ranged between *180 and 290 parts per million*. When earth's atmosphere reached the right levels of gases needed to support the ecosystem upon which life as we know it depends, mammals evolved to eventually include us.

But slowly, insidiously, as we humans embraced burning fossilized sunlight as a source of energy for warmth, transportation, and growing and preparing food, we didn't realize that this blessing came with a curse.

When the industrial age began in 1750, CO_2 parts per million were stable at the *280 PPM level*. By 1824, they were steadily rising, and by 1860 the effects were actually quantifiable. In 1896 it became clear that coal burning could lead to global warming, and by 1938 it was found to already be starting. Research advanced rapidly after 1940, and in 1957 the public was first alerted that fossil fuel burning was "a grandiose scientific experiment" being carried out upon climate.[1]

But by then, a fossil fuel burning car had become part of the American family, my own included. What a blessing our blue and white station wagon turned out to be—my father didn't have to ride two buses to get to work, my mother no longer had to phone in her weekly grocery order, then wait for its delivery, then send back what wasn't right. We didn't have to call on a relative to drive us to the hospital for our latest childhood ailment. And we could take Sunday drives out into the country, as well as spend my father's vacation week making day trips to the beach or the wild animal farm or the amusement park. Life was good!

My partner Milt remembers back when his grandfather plowed the family farm with a team of horses, and then when he shifted to the "horse-power" of a tractor that ran on gasoline after WWII.

By the time of the 1964 World's Fair, atmospheric CO_2 was pushing 315 *ppm*. Did we know? Should I care? By 1971, when I was driving a mid-size station wagon so that my own two small daughters could sleep in the back on long trips, Exxon oil company's own scientists had confirmed the emerging consensus that fossil fuel emissions could pose risks for society,[2] and began "exploring the extent of the risks."

Meanwhile, I think our burnt orange and white station wagon averaged 8 mpg, but gas was inexpensive and not an issue. Then in 1973 the Organization of Petroleum Exporting Countries (OPEC) disrupted oil supplies in response to an Israeli war backed by American weapons. The resulting gasoline shortage had us waiting in long lines for $3 worth of gas at a time, odd-number license plates on odd days, even numbered ones on even days. A fifty-five mile an hour speed limit was imposed to further conserve gas. The U.S. economy went into a recession. and the American public began to pay attention to oil.

In July 1977, a senior scientist in Exxon's Research & Engineering division warned company executives of the danger of atmospheric carbon dioxide increases from the burning of fossil fuels,[2] and reported that there was general scientific agreement at that time that the burning of fossil fuels was the most likely manner in which mankind was influencing global climate change.[2] CO_2 levels had reached 339 *parts per million.*

I was out of the country by then, accompanying my U.S. Army officer husband on a tour of duty with NATO in Germany. While we were stationed in Europe, I worked on a degree in Peace Studies, trying to create options to wars that put my friends and neighbors into harm's way. Geopolitically, we were caught up in the Cold War between two nuclear super powers. But while I was attending a seminar at the U.S. Army Russia Institute in Garmisch, Germany and listening to civilian and military experts coolly discuss Mutually Assured Destruction, I had no idea that another potential Armageddon was already unfolding.

A 1979 paper presented to President Carter signed by four distinguished scientists warned that the time for implementing policies was passing; global warming would probably be conspicuous within 20 years, and enlightened policies managing fossil fuels could delay or avoid the changes.[3] At that point the CO_2 in the atmosphere had increased to *339 ppm*. Yet instead of dealing with this reality by finding and focusing on renewable non-carbon-emitting energy, the newly formed Department of Energy began promoting a massive program of synthetic fuels to be made from coal, tar sands, and oil shale.[3] Because these synthetic fuels would produce more climate-altering gases than most other energy technologies, the government promptly produced a memo to downplay the growing climate concerns.

Returning state-side with a diesel-fueled vehicle we named Eliot for Pete's dragon in a favorite children's story, I was oblivious that public awareness of the greenhouse effect of burning fossil fuels was increasing, or that conservative reaction was gearing up to deny any environmental concerns that could lead to governmental regulations and threaten profits.

By the 1981 Presidency of Ronald Reagan, global warming had become a political issue. To implement spending cuts on climate-related environmental research and to stop funding for CO_2 monitoring, Reagan appointed an Energy Secretary who claimed that there was no real global warming problem. *By then CO_2 parts per million were topping 340 and global temperatures had risen a half a degree.*

Congressman Al Gore, who had once studied under a leading climate scientist, joined others in arranging for congressional hearings from 1981 onwards. These hearings gained enough public attention and concern to reduce the funding cuts for atmospheric research...for the time being.

But a polarized political debate developed. As the research became more solid, contrarians began claiming that increases in CO_2 should be encouraged and not suppressed, because CO_2 is good for plants. In 1983 an Environmental Protection Agency report said global warming was "not a theoretical problem but a threat whose effects will be felt within a few years," with potentially "catastrophic" consequences.[4] When the Reagan administration reacted by calling the report alarmist, the dispute got wide news coverage. CO_2 emissions were *340 ppm*.

I was tooling around D.C. in my diesel dragon that sounded like a tank, smelled like a tank, and drove like a tank. The changing climate was *not* on my radar screen. Nor, apparently, was I paying attention to the summer droughts and heat waves in full swing when NASA scientist James Hansen testified in a Congressional hearing on June 23, 1988. He stated with high confidence that long term warming was underway, with severe warming probable within the next fifty years, likely causing storms and floods.[5] CO_2 emissions had reached *351 ppm*.

It's not that I was politically absent and abstinent. In fact, utilizing my new M.A. in Peace Studies in the 1980s, I was part of a grassroots campaign to establish a federally funded institute for peace. I rode to the Senate committee hearings as far as the Pentagon with my husband's carpool, encouraged by my military neighbors to open up non-violent options to the wars they had to fight. Ironically, when the bill finally passed Congress, it was attached to the military budget and signed by Reagan. And I had learned a valuable lesson about how our government really works, in contrast to what I'd been taught in high school civics class.

But if there was increased media attention on the scientific community's broad consensus that the climate was warming, that human activity was the likely primary cause, and that there would be significant consequences if the warming trend were not curbed, I must have missed it. Apparently, these facts did encourage some discussion about and interest in passing new laws concerning environmental regulations that were simultaneously and vigorously opposed by the fossil fuel industry.

From 1989 onwards, with CO_2 *ppm pushing well into the 350s* and global temperatures inching towards *one full degree of warming*, the fossil fuel industry funded organizations "to spread doubt among the public in a strategy already developed by the tobacco industry."[6]

In fact, some of the tobacco industry's "scientists" became vocal against the prevailing climate science findings, and, supported by and in support of conservative interests, became politically involved; instead of publishing their opinions in peer-reviewed science journals, they spoke directly to the public through their articles, books, and the press.

This became a period when legitimate skepticism about basic aspects of climate science was no longer justified. Those spreading mistrust about this issue became "deniers." As their arguments were increasingly refuted by the scientific community with new data, these deniers turned to political arguments, made personal attacks on the reputation of mainstream scientists, and promoted ideas of a global warming conspiracy.

With the 1989 fall of communism, the attention of conservative think tanks turned from the "red scare" to the "green scare," which they saw as a threat to their goals of private property, free trade market economies, and global capitalism.[5] Into the 1990s, conservative think tanks launched campaigns against increased regulations on environmental issues from acid rain to ozone depletion, second-hand smoke, and the dangers of DDT, using the argument "that the science was too uncertain to justify any government intervention."

This strategy would continue for the next two decades, influencing public perception and discourse by shifting it from the science and data of climate change to a discussion of politics and the so-called controversy. For instance, an advertising campaign funded by the coal industry sought to reposition global warming as "theory" rather than "fact."

Yet worldwide concern over global warming had prompted the creation of an Intergovernmental Panel on Climate Change assessment that resulted in the UN Framework Convention on Climate Change, signed by all countries in 1985. The IPCC began as a small group of scientists mandated to assess scientific information relevant to human-induced climate change, its impact, and the options for adaptation and mitigation. It serves as an autonomous intergovernmental body in which scientists take part both as experts on the science and representatives of their governments, produce reports which have the firm backing of all the leading scientists researching the topic, and which then have to gain consensus agreement from every one of the participating governments before its reports are shared with the public. And because IPCC climate change assessments include input from scientists from all the world regions as well as policymakers representing all the world governments, endorsement of its findings are reliable, incontrovertible, and widely quoted.[7]

At the Earth Summit in Rio de Janeiro in 1992, 154 nations signed U.N. Framework Convention on Climate Change that called for governments to reduce atmospheric concentrations of greenhouse gases with the goal of "preventing dangerous anthropogenic interference with Earth's climate system." But President George H. W. Bush became the loudest voice in the room calling for mandatory emissions cuts to be replaced by voluntary ones, and only signed once this revision was made.

He cited "scientific uncertainty" and economic risk, and declared "the American way of life is not up for negotiation."[8]

Meanwhile, the 1990s found me in seminary, where I celebrated my 50th birthday as a first-year student in 1994. After living in a nine-room, three-bathroom home, I'd moved into one small dorm room, with a community bathroom down the hall. I had no phone and no car, both by choice: I needed silence and solitude to discern what was next in my life.

At the end of the first year, I went home for the summer to end my thirty-year marriage and fully move out of the family home. After selling my relatively new luxury car because it felt obscene to keep it, I returned to Berkeley and made my way to the nearby Saturn dealership.

A spin-off of General Motors, this new company promised a different kind of buying experience, a commitment it more than lived up to—I was treated as an intelligent woman, capable of deciding what best fit my needs, wants, and budget. As an added bonus, I got to finance my first solely owned car and establish a credit history for my new single life.

I promptly named my new car Columba, short for columbine—Latin for dove—the wildflower that had become a metaphor for my new life. Columba was a pale plum color, and could get 28 mpg in the city, 37 on the highway. Meanwhile atmospheric CO_2 had reached *360 ppm.*

After the previous year's walking as far as my legs and available time would let me, I now had access to hiking trails up in the Berkeley Hills, and Marin County and the Pacific Coast highway were barely an hour away. My car became my modern monastic cell as it carried me across the Northern Rockies to and from my chaplaincy training summer in Billings, Montana. After graduation, I drove clear across the country for my parish ministry internship in Massachusetts. I was finally back to where I'd started from as a child, and, wanting the chance to spend time with my family of origin as my parents aged, I took a part-time ministry position in a small church south of Boston. Columba was terrific in the traffic congestion while trying to get through the city to my ailing father's bedside in the late 1990s, and then to conduct his funeral.

CO$_2$ levels were steadily climbing towards *370 ppm* and I was still oblivious of this growing problem that I was perpetuating! How could that possibly be? In 1998 the American Petroleum Institute wrote a proposal intended to recruit scientists to convince politicians, the media, and the public that climate science was too uncertain to be taken seriously.[5] This proposal included a five million dollar multi-point strategy to "maximize the impact of scientific views consistent with *ours* upon Congress, the media, and other key audiences." Their goal was to raise questions and undercut prevailing scientific wisdom.

By the beginning of the 2000s, the efforts by climate change denial groups were recognized as an organized campaign. Taking a page from the tobacco campaign, these propagandists began receiving funding from oil companies. ExxonMobil led in corporate donations to these think tanks, and between 1998 and 2014 gave nearly 31 million dollars to groups that would deliberately spread climate misinformation.[9]

The ideologically conservative Koch brothers, with their massive petrochemical business interests, donated more than $100 million from 1997 onward to 84 groups promulgating climate denial, all shielded from public scrutiny through financial vehicles known as Donors' Trusts.

As the money flowed through this dubious network over the decades, its misinformation strategies passed like a baton to a shifting array of coalitions and initiatives that protected fossil fuel interests in the climate debate. Some groups produced reports that cast doubt on the accumulating evidence of manmade climate change, and others amplified the alternative findings. Think tanks in the network held conferences, sponsored panels, wrote op-eds and letters, and created an echo chamber loud enough to command equal time in the mainstream media.[9]

In 2000, environmentally aware and climate savvy Vice President Al Gore ran for president. When the Supreme Court ruled on the contested election results and handed the presidency to oilman George W. Bush, I was in the middle of an interim ministry year in Las Vegas, Nevada.

With my little church back in New England barely able to pay me for part-time work, and aching to do full-time ministry, I had put my name into the interim ministry pool, a group of ministers who undergo special training to serve congregations "in transition" between ministers. But I was surprised and shocked to find myself "banished" to the desert, where brown replaced familiar, comforting green, and trees were scarce.

Because I stay grounded by bonding with my natural surroundings, I had to learn to see and appreciate the desert's gifts. By mid-year I was smitten, so said yes to the religion reporter from the local paper that wanted to interview me about religion and environmental awareness. Apparently, she had struck out with other local congregations, but my faith tradition includes Emerson and the other Transcendentalists for whom nature is a primary scripture, so I had the theological backing for what the reporter was looking for.

When the article appeared, complete with photos shot during a worship service, the other half of it included an interview with Josh Abbey, representing the Jewish tradition. He stated that his infamous father would have written off Las Vegas long ago for its ecological transgressions. It was at that moment in time that I consciously moved from being a nature lover into becoming an environmental activist.

When a congregant connected with the Sierra Club asked me to do a workshop with the local members who were feeling totally defeated by what was happening in D.C. I said yes. Two months after George W. Bush was sworn in, he renounced the Kyoto Protocol, the international treaty which extended the 1992 UN Framework on Climate Change that committed state parties to reduce greenhouse gas emissions.

Then, under Vice President Cheney, our national energy policy was rewritten behind closed doors to include the Halliburton loophole, thus exempting his old company from clean water regulations, and kick-starting the hydraulic fracturing frenzy to ensure access to the "bottom of the barrel" fossil fuels that would emit even more CO_2, plus release methane, an even more potent greenhouse gas.[10] With CO_2 emissions climbing through *370s ppm*, we were clearly going in the wrong direction!

By 2002, I'd begun to be concerned about my own carbon emissions. If I were going to do interims for my ministry career, I would be driving back and forth across the continent over the next several years. My sweet Saturn was fuel-efficient, but I had heard that the company was developing a hybrid vehicle that would run on battery as well as gasoline.

As I eagerly waited for this option, I did *not* know that its parent company, General Motors, had developed, produced, and leased electric vehicles (the EV1) between 1996 and 1999. It was the first mass-produced electric car in the modern era by a major automaker, and the only electric passenger car to be marketed under the GM brand name. The decision to mass-produce an electric car came on the heels of a mandate by the California Air Resources Board that required the sale of zero-emissions vehicles from the seven major automakers if they were to continue to sell their vehicles in California.

These EV1s were made available via lease-only agreements to residents in selected western cities and could be serviced only at designated dealerships. Consumer reaction to the electric cars was so positive that GM grew worried that these cars wouldn't prove profitable enough. After all, without gas motors there is little need for maintenance, or demand for gasoline. They rounded up the leased cars, refused to sell them to the lessees who wanted to purchase them outright, and crushed them.

Meanwhile, the car manufacturers litigated the CARB requirement for the wiggle room to make and sell super low emission vehicles and hybrids instead of going the all-electric route. When the EV1 program was discontinued in 2002, the oil companies, along with the oil-drenched executive branch of the federal government, had played a huge role.[11]

By 2002 I was in my third interim ministry, in Colorado Springs, Colorado. I'd spent my second interim year with a congregation in Vermont that was steeped in environmental awareness: I went there because I had so much to learn! But before that year barely started, 9/11 happened.

Our national response to the horror was to bomb Afghanistan; that it is an oil, gas, and coal-rich country was probably just coincidental in our determination to punish someone for our national tragedy. Yet the decision to bomb Iraq in the spring of 2003 was different. It was widely suspected that we were after their oil, and demonstrations were held worldwide, including in the community where I was living.

Months before that unpopular invasion, I had given up on waiting for Saturn to market its hybrid, suspecting but not knowing for sure that GM had opted for making the more profitable Hummer instead.

Wading through the grief of giving up my faithful companion, I traded Columba in for a Toyota Prius that wanted to be named Gaia. She was the soft aqua shade of earth and water and sky combined, and promised to get 50 mpg in town, somewhat less on the highway, with super-low CO_2 emissions. The reverse of gas burning cars, the hybrid gets better mileage when constant braking at stoplights recharges its lithium battery. This takes some getting used to: whenever I stopped at a light, and the engine went silent, I worried it had shut down completely.

I'd checked out the Honda hybrid as well, but it didn't have the super low emissions rating that Toyota's did. The early Prius I owned still had the Corolla body, not the unique one seen on our roads today. I would have preferred buying American had there been any way to do so. Only U.S. dealerships whose mechanics had received special training in Japan were allowed to carry the Prius. I came to feel like a pioneer, as glitches worked themselves out—or not. But these annoyances were worth it to me to be doing the right thing. I was a minister, after all!

My commitment was unexpectedly confirmed at a stoplight in Santé Fe, New Mexico, as I headed back to Colorado after an annual interim ministers meeting. The driver of the pickup truck beside me rolled down his window and shouted, "What mileage do you get in that thing?" When I told him, he whistled, then declared, "I'm gonna get me one of those. Let them Arabs keep their damn oil!"

That we the people were catching on became hopeful during my next interim ministry, in Flagstaff, where it seemed that every other car was a Prius in this progressive college town in northern Arizona, just 70 miles from the south rim of the Grand Canyon. I agreed to stay on for a second year, part-time, so I could have sabbatical time up at the Canyon.

The second autumn I was there found me standing at the checkout counter at the nearby Albertson's, leafing through the September 2004 issue of the *National Geographic*. Its title, "Global Warning, Bulletins from a Warmer World," had caught my attention, and the inside note from the editor piqued my interest enough to buy the copy on the spot.[12]

Essentially, he claimed that, after a decade as editor in chief, he had a good idea what articles would provoke a lot of angry letters, and even terminate memberships, yet he wouldn't be able to look at himself in the mirror if he didn't bring readers the "biggest story in geology today."

Seventy-four pages of stunning photographs from across the planet *showed* rising sea levels, melting glaciers, thawing permafrost, increasing wildfires, lingering droughts, shrinking lakes, changing habitats and migration habits, bleaching coral reefs, invading exotic species, disappearing amphibians, eroding coastlines, and ice shelves collapsing. Then he challenged even those who don't believe the Earth is getting warmer and that human behavior is a contributing factor to take a look at the hard truth as scientists were seeing it. CO_2 levels had passed *380 ppm* by then. Would the public finally wake up?

At that exact moment in our country's history, the billionaire oil-baron Koch brothers began building an assembly line to manufacture political change that included think tanks which produced papers, advocacy groups that pushed for policies, and PACs that donated money to candidates. By putting these all together, the Kochs were able to push back against doing anything about climate change on those three fronts all at once. "You get papers that look like they're real scientific opinions doubting that climate change is real, you get advocacy groups saying we can't afford to do anything about it, and you get candidates who are told if you want to get money from Koch donors you have to sign a pledge saying that, if elected, you will do nothing about climate change that requires spending any money on the problem."[13]

No wonder that when I went back east for my fifth interim ministry, this time in Charleston, South Carolina, hybrids were few and far between, and I could count them on one hand while driving route 95 for two days from the southeast to the northeast to visit my ailing mother.

In fact, the neighbor next door to my rented townhouse on the Ashley River drove a honking huge SUV, mainly because her tax advisor recommended it. As a real estate agent, she would get a tax advantage that would be foolish to refuse: her expensive sports utility vehicle could be depreciated more rapidly than anything under 6000 pounds. What was its gas mileage and carbon footprint? Why should/would she care?

Then Hurricane Katrina flooded New Orleans! Surely *now* people would pay attention to what was coming to pass as we wantonly dump CO_2 into the atmosphere, warm the earth and alter the climate, thus creating such a superstorm surge. But the national media never once connected those dots during their massive coverage of the unfolding disaster in Louisiana.

Why not! What was going on here? Was there an unspoken conspiracy between news outlets and their fossil fuel industry and car company sponsors? How could that be happening in a nation built upon the foundations of reason, logic, and science?! After all, CO_2 had just surpassed 380 ppm! Yet when astronaut Sheila Collins returned from space that fall, and felt compelled to report what she had been seeing over time—shrinking glaciers, expanding deserts, disappearing forests—no one paid attention. The major news story on that day was about the risk of driving while wearing flip-flops.

After a frustrating year being mired in obliviousness, I was off to Maine for a final interim ministry year. It was 2006, and *An Inconvenient Truth* was being made available to selected congregations through Interfaith Power and Light. I borrowed the video to show at our church and share with the Yarmouth community. In the graphic film he had created and produced, Al Gore re-emerged on the public scene to set out the science of climate change and discuss concerns around global warming. Even before he lost the 2000 presidential election, Gore had been internationally involved in trying to deal with the greatest challenge to ever face humanity. Now, he was about to be awarded the 2007 Nobel Peace Prize jointly with the U.N. Intergovernmental

Panel on Climate Change (IPCC) for their efforts to build up and disseminate greater knowledge about manmade climate change, and to lay the foundations for the measures needed to counteract such change.'[14]

Also in 2007, a *Newsweek* cover story reported, "the denial machine is running at full throttle." This well-coordinated, well-funded campaign by contrarian scientists, free-market think tanks, and industry had "created a paralyzing fog of doubt around climate change."[4] CO_2 emissions were nearing *390 parts per million.*

In 2007, I finished up my interim ministry career and took a settled position, back in Las Vegas, where it had first begun. I desperately needed to stay put in one place, put down roots, buy a home of my own, and *not* move every summer. The desert spoke to me as no other place had, and so I settled there. By 2008, the Bush presidency ended and the Obama one began, and I felt we could take a deep breath. Surely now our government would see the light and do the right thing.

But in 2010, a Supreme Court decision removed caps on corporate and nonprofit political donations and opened the floodgates on campaign spending. Then billionaires such as the Kochs moved millions of dollars to support the rise of the Tea Party movement and ultra-conservative candidates who used climate denial as a bedrock of party orthodoxy.[10]

Soon, few Republicans running for federal office would admit to accepting the reality of manmade climate change. CO_2 was approaching *400 ppm*.

By 2010, I was in a partnered relationship with a retired aerospace engineer who was as concerned and committed as I was about the fate of the warming world. And we would soon be driving a Chevy Volt, GM's plug-in electric car, and powering it by solar panels on the roofs of our individual homes, mine in Vegas, his in Denver.

Ironically, the CEO of General Motors had admitted that the biggest mistake he ever made was killing the EV1 and failing to direct more resources to electrics and hybrids after having such an early lead on this technology. Was the Volt my Saturn revamped?!"

We named it Emerson, for the Transcendentalist writer and philosopher, because we saw it as being, like Ralph Waldo, quietly subversive on behalf of the planet. We called its GPS system Naggie Maggie after Margaret Fuller, a persistent stone in Emerson's shoe.

We were about to become climate activists.

TWO. ORANGE
ACTIVISM

"The insanity of human destructiveness may be matched by an older grace and intelligence that is fastening us together in ways we have never before seen or imagined."

—Paul Hawken

AT WHAT POINT DO PEOPLE PUSH BACK? When does each of us decide that enough is enough? And what does it take to take a stand, then stand up and proclaim: this will not stand! I suspect that the break-through point is as unique as the individual who reaches it! Sometimes such personal moments become public movements.

Take Occupy Wall Street. When the housing market tanked because the banks were creating toxic mortgages and then betting against them, ordinary people were affected. Retirement nest eggs disintegrated, and family homes were foreclosed upon. As I watched the value of my newly acquired home drop to half of what I still owed on it, I was glad and grateful that I hadn't let my real estate agent sell me one of the properties she'd insisted I *qualified* for, based on my projected income. Or worse, fall for the adjustable mortgage scheme that let buyers purchase homes they couldn't afford, only to have their mortgage payments jacked up higher, later.

Instead, I settled for what I would be able to afford even without my salary, something with a mortgage payment that could come out of my partial military pension if/when I retired earlier than expected. I had seen too many colleagues trapped in unhealthy ministries because they couldn't financially afford to leave. I wanted wiggle room. And besides, I much preferred the 1500 square foot duplex I actually bought to the larger detached homes I qualified for.

All of which was beside the point: I too was outraged that the Wall Street entrenched banks, being "too big to fail," were bailed out by the government, while the hard-working other 99%, as homeowners, were thrown out on main street, and found myself joining up with Occupy Las Vegas that set up camp on an authorized vacant city space.

While I never actually slept in the tent I put up for two local college students, I did spend a fair amount of time at the encampment, marveling at the juxtaposition of the triangle shape of my 4 person tent silhouetted against the skyline of the Strip...how it perfectly mirrored the pyramid shape of the Luxor Hotel. What an apt reflection of the casinos that the banks had become. Yet their fantasy bubble had burst in *our* faces.

One of the books I donated to the library tent was a copy of *The Power of the People*. This was a text I'd studied while working on my M.A. in Peace Studies; in turn, I used it in classes I taught on peace and justice issues. A definitive history of *Active Nonviolence in the United States* that actually starts back *before* we became an independent nation, this book depicts the movingly illustrated alternative narrative of who we are. For while the majority will forever mindlessly follow the status quo, there has always been a segment who questioned it and, when things didn't make sense or feel right, sought to change it. In fact, there had been a plethora of people pushing back even before Thoreau inspired subsequent generations to take up the right and duty of civil disobedience. It was being taxed to support the war with Mexico he didn't agree with that pushed Thoreau over the edge and landed him in the Concord jail overnight.

But usually *last straws* come about after a long, frustrating struggle to work through the system to right perceived wrongs. Nearly all the things we now take for granted were fought for by someone at some point, from civil rights to worker rights to environmental rights.

Take women's right to vote: it took some seventy-plus years of lecturing, rallying, lobbying, and finally chaining themselves to the fence of the Wilson White House, being hauled off, harassed, imprisoned, and force-fed before the suffragists finally prevailed.

Unionizers were brutalized and murdered before workers obtained decent hours and pay, plus safer working conditions. The fight for civil rights still conjures up snarling dogs and water hoses. Environmentalists continue to fight for clean air, clean water, and uncontaminated soil while corporations keep profiting off the commons by dumping their wastes and trashing our earth while expecting the rest of us to clean it up.

It's not too hard to see the thread that connects these movements...*all* pushed back against the main narrative that excluded everyone other than privileged, propertied white men. Call it patriarchy, capitalism, colonialism, imperialism or empire, the entrenched system thrives by exterminating or enslaving "inferior" people and exploiting resource-rich labor and lands.

Yet pushback seems to be encoded in our national DNA. I like to picture this predisposition as a recessive gene that is crouching in our collective unconscious, ready to spring into action when push comes to shove. That shove point is the wild card.

My shove point for climate activism inadvertently came in the spring of 2012, when our Occupy Las Vegas group joined with the Sierra Club, the Center for Biological Diversity, faith groups, and members of the nearby Moapa Paiute tribe who lived beside the coal-fired power plant that provided electricity to Las Vegas, to march together to our local utility headquarters and demand a shift to renewable energy and the closure of the plant that had been poisoning the Paiute people with toxic ash for over 50 years.

It was also my introduction to ALEC, the right-wing American Legislative Exchange Council that drafts legislation for their paid-for politicians to introduce into state and local legislatures: laws that undermine everything I've believed in and worked for.[15] Clearly we were in for a long, fierce fight on many fronts, including around climate change.

Yet I felt torn: by then I had retired from ministry in order to resume my primary passion for writing, and had several manuscripts shape-shifting on my desk that had been nagging at me for years. But face to face with the most critical crisis of our era, or maybe ever, I slowly accepted that climate change had to be a major focus of my life/work at this point in time. Sigh! This much I do know: most serious writers are severe introverts; being immersed in the creative process demands it. Yet while we focus inward, the outer world goes on, making its demands upon us. Could I reconcile both?

One role model for this possibility soon became author-activist Bill McKibben. He began working as a freelance writer at about the same time that climate change first appeared on the public agenda—when the hot summer and wildfires of 1988 *coincided* with the testimony by James Hansen before the Senate Committee on Energy and Natural Resources.

McKibben's first contribution to the debate was a brief list of pieces focused on the topic *Is the World Getting Hotter*.[16] At that moment in time, CO_2 parts per million were at 351, topping the upper safe limit to avoid a climate tipping point.

McKibben's first book, *The End of Nature*, was published in 1989, after being serialized in *The New Yorker*, and became the public's first introduction to the crisis of climate change. When I found and first read the book in 2002, it was already over a decade old, and the introduction had been revised to include what had happened in the meantime. Then in 2006, McKibben helped lead a group of students, all from the college where he taught, on a five-day walk across Vermont to call for action on global warming. In 2007 he founded 350.org and would go on to undertake a *Do the Math* tour, promote divestment from fossil fuels, and create the "keep it in the ground" movement.[17]

I went to hear him when he came to Denver to receive the Rebel With A Cause Award from the Colorado Environmental Coalition in 2011. On that raw rainy night in May, with the Platte River raging beneath the hotel windows, he talked to us about 350.org, the growing worldwide activist network determined to get planetary carbon emissions back to *350 parts per million*. The number was based on information put out by a NASA team in 2008: "any value in the atmosphere greater than 350 ppm is not compatible with the planet on which civilization developed and to which life on earth is adapted." Of course, getting back to that range and staying there "could be the toughest thing humans would ever have to do, but there was no use in complaining about it. It's just physics and chemistry. That's what we have to do."[18]

At that moment in time, CO_2 emissions were nearing 400 *ppm*, making his message as haunting as it was daunting.

Twenty-two years ago, I wrote the first book about climate change and I've gotten to watch it all, and I know that simple persuasion will not do. We need to fight. Now, we need to fight non-violently and with civil disobedience. We will never have as much money as the oil companies so we need a different currency to work in: we need bodies, we need creativity, we need spirit. So far, we've raised the temperature of the planet one degree and that's done all that I've described, it's melted the arctic, it's changed the oceans. The climatologists tell us that unless we act with great speed and courage that one degree will be five degrees before this century is out. And if we do that, then the world that we leave behind will be a ruined world.[19]

Three months later, McKibben wrote that he was hauled away from the gates of the White House, where 65 people had been peacefully sitting in for an hour to urge the president to veto the proposed Keystone XL pipeline, a 1700-mile fuse to the biggest carbon bomb on the continent. [20]

We were there for a simple reason: because it was time. After two decades of scientists gravely explaining to politicians that global warming is by far the biggest crisis our planet has ever faced, and politicians nodding politely (or, in the case of the Tea Party, shaking their heads in disbelief), it was time to actually do something about it that went beyond reading books, attending lectures, lobbying congressmen or writing letters to the editor. With Texas on fire and Vermont drowning under record rainfall, it wasn't just *our* bodies on the line.[20]

Why was the Keystone XL Pipeline such a poster child for the growing climate crisis? Simply this: completing the last leg of the pipeline that indigenous peoples called the "Black Snake" would keep dirty fossil fuels flowing for decades just when we needed to switch over to clean renewable energy sources.

Yet in 2011 the Keystone XL Pipeline was considered a slam-dunk. All that remained to be done was the Environmental Impact Statement, an assessment that was required before the State Department would let it cross over our border from the tar-sands mining site in Canada.

With the public comment period still open, thousands of people, my partner and myself included, poured over the sloppily prepared and utterly biased assessment that had clearly been commissioned by Trans Canada, the pipeline's parent company. This portion of the pipeline would cross the Ogallala Aquifer, source of fresh water for millions. I was among the many who were appalled that the likelihood of spills causing environmental devastation was acceptable in light of unsupportable economic claims, as if ecology and economics weren't from the same root word that meant home. And though thousands of jobs were promised, less than 30 of them would be long-term. Plus, it wouldn't make the U.S. "energy sufficient" because the oil was headed for refineries on the Gulf Coast to be shipped elsewhere. If anything, the price of our gasoline would rise.

Public outrage prompted another assessment, but even this supplemental one was pathetic. Ignoring the consensus among oil executives, bankers, environmentalists, and all who agreed that Keystone XL was central to speeding up the extraction of tar sands, the State Department decided that the project was unlikely to have a significant impact on tar sands development. Seriously? This was unacceptable. The pipeline had to be stopped. Period.

In a 2012 essay for the *New York Times*, James Hansen, (yes, that same NASA climate scientist who had first warned Congress back in 1988), explained that:

by burning the oil trapped in tar sands in Canada, the U.S. would be put dangerously at risk from climate change. The Western United States and the semi-arid region from North Dakota to Texas will develop semi-permanent drought and economic losses would be incalculable. More and more of the Midwest would be a dust bowl. California's Central Valley could no longer be irrigated. Food prices would rise to unprecedented levels.[21]

In February of 2013, McKibben and dozens of other eco-activists, including Hansen, were arrested in front of the White House protesting the Keystone XL pipeline. "We really shouldn't have to be put in handcuffs to stop KXL," McKibben claimed, "but given the amount of money on the other side, we've had to spend our bodies, and we'll probably have to spend them again."[21]

By the summer of 2013, thousands of we the people were signing the Keystone XL Pledge of Resistance: "I pledge, if necessary, to join others in my community, and engage in acts of dignified, peaceful civil disobedience that could result in my arrest in order to send the message to President Obama and his administration that they must reject the Keystone XL Pipeline." My partner and I did likewise, and then we participated in one of the training sessions being held all across the country.

The very first activity at the two-day session was to fill out a pre-printed 8x11 piece of paper that read:

PRESIDENT OBAMA,
I OPPOSE KEYSTONE XL BECAUSE....

Because indeed?

Why would a retired elder risk arrest? What was my bottom line here, a reason succinct enough to fit across the middle of the sheet of paper and be big enough to be seen when held up as part of a protest?

We had recently hiked up to a favorite meadow in the high country and communed with the columbine blooming above the tree line, as we'd been doing every summer. But this year I'd noticed they seemed to be thinning and fading, just as predicted. With a diminished snowpack and an increase of heat, sage will begin to march up the mountain to replace them.

My answer became simply this: THE EARTH IS MY CHURCH.

For me, to pollute the air, poison the soil, contaminate the water, and decimate whole ecosystems for power and profit was pure *evil*. To be complaisant or complicit in evil was to commit a *sin*.

Tragically, my generation, the one that would suffer the least from the effects of a warming world with its wild climate events, was the most culpable: we had come of age and lived our long lives fueled by fossilized energy. Call it atonement (certainly not martyrdom...*that* was not on my bucket list), I had to do whatever I could to stop this trajectory.

It wasn't a choice; it had become a call. How could I *not* step up and take action? After all, I was spending summers in Denver, where the previous year we'd been choking on smoke from the converging plumes of two massive, drought-exacerbated wildfires, one north in Ft. Collins the other south in Colorado Springs. And this summer there were two more fires raging in the state, destroying more homes and bark beetle-infested woodlands. Enough had become enough!

Because our nation has such a long history of civil disobedience, there is a whole body of wisdom around how to best use this tool for change. In the KXL training session, for instance, we learned that every planned action must include a local action leader, a police observer, a police liaison, a volunteer lawyer, plus jail and media support in order to be safe and effective. One cardinal rule for every protest was to ensure that not everyone got arrested: there had to be people remaining on the outside to make sure those inside the prison didn't get lost in the system.

Over the course of the summer of 2013, 10,000 people signed the pledge of resistance and trained up to take action if and when the president approved the Keystone XL pipeline. Later that summer, I acquired a wooden sign to put on my desk that read:

A good friend will come and bail you out of jail…
a true friend will be sitting next to you saying
Damn…that was fun! [22]

But the training session made it very clear that this commitment was a serious matter, with potentially grave consequences. Green was becoming the new Red. Environmentalists were being branded as eco-terrorists for disrupting the status quo and targeted by the Department of Homeland Security. I had suspected as much when, at a community security training session in my Sun City Las Vegas neighborhood, I was horrified to see "environmentalists" on the list of people to watch out for. I left at break, slinking away in my newest Prius, a shiny white vehicle I had named Starr.

Now here I was five years later, ready to do whatever was necessary. That autumn of 2013 was a tense time as we awaited the president's decision on the KXL pipeline. I wrote him a letter out of my own angst; many others did likewise. He had to know that so many of his fellow Americans were ready to protest if and when he said yes to Trans Canada.

In November, at the recommendation of the State Department under John Kerry, President Obama said NO. In its first action of 2014, the new Congress tried to force the president to approve the pipeline but did not have the votes to override a veto. Although for the rest of Obama's term the Keystone XL pipeline was "off the table," it lay dormant beneath it, like a poisonous snake awaiting resurrection.

But the climate movement had been mobilized!

It was as if something vital had awakened and was manifesting in well-known and unknown people who were all pushing back. Connecting our stories and struggles, we were building a net for safety, for strength, of inclusion, of support, for the greater work of saving the world from being pillaged and plundered for political power and corporate profit. We were combining forces, becoming a community of resistance as if the planet's very own immune system was revving itself up. And we were connecting not just actually but virtually, via the internet…and, more often than not, both.

I for one had just joined Facebook by that spring of 2014, after my newest book was published and my written contract called for a social media presence and platform. On my seventieth birthday I posted a photo taken as I'd sat among the columbine the previous summer. In the caption I made a birthday wish, "that within my lifetime we the people take our planet back from the fossil fuel industry that's changing our climate by pouring carbon back into the atmosphere.… mindful that the temperatures at the 2014 Winter Olympics hovered around 60 degrees, prompting over one hundred athletes to sign a letter demanding that world governments start dealing with the climate crisis NOW."

I suggested that if my "friends" hadn't heard much about this it was probably because British Petroleum was a major sponsor of media coverage of the Olympics. This felt crazy-making to me, so my birthday candles glowed in honor of the growing number of organizations trying to break this cycle of insanity, McKibben's 350.org among them.

Also, because I was on Facebook, I was able to virtually join in a Great Climate March as it set out in a rainstorm from Los Angeles that March. This group included youth and elders and every age in between, some 35-50 ordinary folks, each with a unique story of why they were willing to walk across the entire continent to, in the words of their mission statement: "change the heart and mind of the American people, our elected leaders, and people across the world to act now to address the climate crisis. Marchers and many people we meet along the way continue to push back against climate change through both direct action and political action. We are marching, and invite you to join us in the fight for our lives!" [23]

While they were trekking through the Mojave Desert, I'd emailed pertinent pages from my book *Chewing Sand, an Eco-Spiritual Taste of the Mojave Desert*, to an active Facebook poster named John, who then let me know he had taped my essays up on the walls of the port-a-potty. As a writer, I felt strangely honored and pleased.

I caught up with the marchers in person when they came through Denver in mid-June. They were just setting up camp in the parking lot of a church still several miles south of the city. I got to talk with many of the participants over supper and then read from my book by flashlight to a small group still awake after a long day on the road. I came back during breakfast to share more essays with a young woman who requested it. Then I dried dishes washed in a soapy bucket and rinsed in water treated with chlorine, awed by the commitment of these teens and adults and elders.

At the send-off closing circle, I wished them well with a communion of tiny plastic dinosaurs, because the Denver Basin they were walking through represented the deep time during which the fossil fuels were laid down, with gas extraction now being resisted by communities along the Front Range. In fact, due west of Denver rose Dinosaur Ridge, where some of the earliest excavations of this extinct species had taken place. While the dinosaurs didn't see their fateful asteroid coming, we humans do, and I thanked the marchers for trying to wake us up before our own extinction became inevitable!

Later in the day, my partner Milt and I joined the group in a rally at the Colorado State Capital. After speeches and street theatre presentations, we all marched the entire length of 16th Street to an informational lecture at the Alliance

Center, then schlepped back across town to a vegan restaurant that had prepared a buffet dinner. Grouped around tables to discuss and then report out on pertinent topics around climate change, the marchers and members of the greater Denver community shared their angst and hope and commitment to continue calling awareness to and demanding action on this looming crisis.

The Great Climate March expected to connect up with the People's Climate March scheduled for September 21st in New York City. While over three hundred thousand people showed up to take part there, companion events took place worldwide, with nearly 600,000 people involved in actions, symposiums, and presentations leading up to the U.N. Climate Summit of world leaders set for September 23rd. Unwilling to add to my personal carbon footprint by flying to NYC, I instead co-led a climate awareness workshop in Las Vegas, where I had just returned to spend the winter months. Attended by a dozen people of all ages, the event concluded with a communion of little plastic dinosaurs as a symbol of commitment to work against the coming next mass extinction.

The People's Climate March itself was a response, not a protest, and was endorsed by over 1500 organizations, including many international and national unions, churches, schools and communities concerned about environmental justice as a climate change issue.[24]

As "the largest single event on climate that has been organized to date....one so large and diverse that it cannot be ignored" according to its founder Bill McKibben, the People's Climate March was billed as "an invitation to change everything," referring to a newly popular book by Naomi Klein: *This Changes Everything: Capitalism and the Climate.*

Milt and I got a copy right away and began reading up on how the climate crisis could become the catalyst for change from the existing broken economic and political system into one that could bring forth a shared, sane, and sustainable world.

We started 2015 determined to process Klein's book with interested others, and co-created a group that within weeks was tabling at the local farmer's market to urge divestment from fossil fuels, an effort to stop feeding the beast that was devouring our future! If indeed in our capitalist system money is the measure of all things and more is never enough, then we can begin the great turning to something new by dismantling the old, starting with keeping fossil fuels in the ground and stranding its assets.

Although there were only six of us, ranging in age across four decades, we outfitted ourselves with banners and canopy, table and folding chairs, and soon grew into what we would come to call LVCA: Las Vegas Climate Action. We started a Facebook page with this mission: "Because knowledge is power, we are coming together to study Naomi

Klein's *This Changes Everything: Capitalism vs. the Climate.* The initial six of us represented a wide range of ages, experiences, interests, and efforts on behalf of the planet (writers, environmentalist, nurse, engineer, doctor, clergy). While empowering ourselves and resourcing one another, we're also already planning our first collective action. Stay tuned!"

We showed up at farmers' markets and outdoor events such as Earth Day with our flyers about divestment, and always with a basket filled with little plastic dinosaurs as reminders of what we were trying to do. Our efforts at divestment eventually included a workshop on divesting one's personal finances from fossil fuels and then reinvesting them in renewable energy, not just for one's own home but for affiliated churches and universities as well. When there was a worldwide march before the UN climate summit in Paris (COP21) in December of 2015, we took part in the Las Vegas rally and marched the length of Fremont Street, complete with divestment signs and a "mini-float" Milt had constructed that featured a large plastic Tyrannosaurus Rex.

The whole world held its collective breath as COP21 got underway. After twenty years of receiving reports from the expert scientists studying climate change yet failing to act on the increasing evidence of its human causes, governments sent their official negotiators, as usual. And, as usual, the U.S. delegation was infused with the influence of the fossil fuel industry, if not as an actual seat at the table, then with a nose under the tent flap held up for it by the U.S. team.

This was the make or break year, and everyone knew it. Official observers, Non Governmental Organizations (NGOs), and ordinary people also converged on Paris to urge governments to finally *do* something before it was too late. And that something would have to be more than merely limiting the emissions of the major greenhouse gas polluting nations: a green fund needed to be set up to help developing countries leap into a clean energy future, and there had to be planning to mitigate the effects already being suffered by those who'd contributed least to the problem.

Those of us who weren't physically attending found creative ways to participate. One way was to write letters to the future that would become a virtual record of why we all wanted this climate summit to matter. Excerpts from mine:

> *If you are reading this, dear ones, it means that humanity matured in time to not go the way of the dinosaurs.*
>
> *Please know this did not come to pass without massive effort on the part of people you may never know about.*
>
> *Hundreds upon thousands of us woke up and worked frantically to avert the "asteroid" of our own care-less creation.*
>
> *Each in our own way, we faced our horror at what was happening to the planet, reached down through our*

grief into layers of gratitude for all life, and touched-in with the cosmic grace at the core of being; then we transformed our sense of hopelessness into activism on behalf of all that we cherished.

Personal responses (rooftop solar, electric cars) quickly became political response—ability as we strove to move public policy away from extracting and burning greenhouse-gas emitting fossilized sunlight. But we not only envisioned a more sustainable future, we relentlessly struggled to create it.

Yet as we began to change how we fed ourselves, heated our homes, moved around, invested, communicated, and lived more simply, the consumerist system we were embedded in undermined our efforts by denying and outright lying about the problem. Anger that could have destroyed us empowered us instead! Rather than remaining isolated, we joined together.

Children and elders, students and retirees, scientists and clergy, mothers and fathers, authors and activists, indigenous peoples and environmentalists, we the people studied, testified, organized, formed coalitions, lobbied, wrote letters, petitioned, demonstrated, marched, prayed, preached, tweeted, voted, sued, protested, got arrested…and escaped into Nature to cry with the thinning columbine. Then we came back and kept at it.

The 2015 Paris Climate Talks should/could/would have been the fruition of our concerted efforts to secure and ensure a viable future. The rest, as YOU will see and say, is history.

For the twelve days that COP21 met, I posted a daily reflection, accompanied by an inspiring illustration that Milt provided, for the UU Ministry for Earth website. The daily ritual felt sacred and connecting.

As COP21 came to a close, with everyone celebrating and congratulating themselves on what HAD been achieved, those of us who'd been following the problem and crunching the numbers couldn't shake the sense that it was too late and too little. Adding up the reduced emissions that each nation had committed to locked us into a more than 2-degree warmer world by 2050, even though everyone knew that to keep the planet habitable we couldn't exceed 1.5 degrees. Plus, the agreement was non-binding (a concession to the fossil fuel industry?). Each nation would be on its honor to live up to its promised emission reduction.

Our country's commitment was predicated on a Clean Power Plan that focused on reducing emissions from coal-burning power plants as well as increasing the use of renewable energy, plus energy conservation. Unfortunately, it also called for substituting natural gas for coal-powered electricity, ignoring the reality that the methane leakage from hydraulic fracturing wells was an even more potent greenhouse gas.

Yet even as most states prepared to meet the requirements, the attorney generals of several states whose economies were dependent on coal mining and burning sued to block action on the Clean Power Plan. The fossil fuel industry just kept on keeping on.

And in a little known area of North Dakota, a group of indigenous people began setting up the Sacred Stone Camp in the path of yet another pipeline that was projected to run from the state's oil fields to southern Illinois, crossing beneath the Missouri and Mississippi Rivers, as well as under part of Lake Oahe near the Standing Rock Sioux Indian Reservation. Calling themselves "Water Protectors" rather than protesters, tribal elders and members established a center for cultural preservation and spiritual resistance to the pipeline.

Over the summer, the camp grew to include thousands of people from across the country: many other tribes, veterans, clergy and chaplains of all faiths, medics, nurses, environmental justice and climate activists, and ordinary outraged citizens all determined to block the *Black Snake's* inevitable threat to the water supply of thousands of people and ecosystems there and downstream, as well as its illegal trespass of indigenous lands. In fact, routing the pipeline near Bismarck had been rejected because of its proximity to municipal water sources, residential areas, and road, wetland,

and waterway crossings. The Army Corps of Engineers had conducted a limited review of the route and found no significant impact. No formal Environmental Impact Assessment had been ordered because a special permit process that treated the pipeline as a series of small construction sites had created an exemption. [25]

After months of the stand-off in severe winter conditions, during which protectors were brutalized by tear gas and attack dogs, water cannons sprayed in freezing weather, rubber bullets fired by hired security guards, arrests of over 600 non-violent protesters, and the harassment of independent media journalists, the Corps of Engineers denied the easement construction of the pipeline under the Missouri River, and an EIS was ordered.

Then late in 2016, a minority of Americans elected a climate denier for president. Never mind that, even as he ranted on the campaign trail that climate change is a hoax perpetrated by the Chinese to destroy the American economy, he was reportedly building walls around his own coastal real estate properties to hold back the rising sea level expected from the melting of polar ice caps caused by global warming.

Among his first actions upon taking office was to greenlight both Keystone XL and Dakota Access Pipelines, claiming that building these pipelines was a no-brainer. Besides, he'd heard NO opposition. Plus, the EIS and water protective regulations were just too burdensome. He did not disclose his financial investment in KXL, with its conflict of interest.

Meanwhile, he stacked his cabinet with executives and lobbyists from, and political apologists for, the fossil fuel industry. Very quickly, every even minimal gain that had been so painstakingly made was systematically dismantled as he withdrew us from the Paris Climate Agreement, dismantled the Clean Power Plan, expunged any reference to the changing climate with its resultant warming, rising, acidifying, etc. from government websites. Clean water and air standards were gutted, public lands appropriated and leased for private oil and gas exploration and exploitation, the waters off the coasts opened for oil drilling, and the Arctic National Wildlife Refuge opened up in the national push to become (dirty) energy dominant because "America's wealth is under its feet."

And, in reaction to the #NODAPL resistance, fossil fuel funded ALEC (the American Legislative Exchange Council) drafted laws that quickly passed in multiple states, criminalizing non-violent civil disobedience.

The fossil fuel industry had completely hijacked our democracy!

THREE. YELLOW

SUNLIGHT

"The sun, with all those planets revolving around it
and dependent on it, can still ripen a bunch of grapes
as if it had nothing else in the universe to do"

—Galileo

IT ALL BEGAN INNOCENTLY ENOUGH, with an offhand remark by a relative: a popular chain store was selling solar-powered Christmas lights. Putting up outdoor lights for the winter solstice had never appealed to me, what with its tangled wires and the requirement to go out in the cold to turn them on and then off. I knew-who-would be doing that, in my pajamas!

But Milt and I were about to celebrate our first Christmas season as a couple, and the urge to do something different and special won me over. On a whim we went to check out solar Christmas lights at a local big-box store. The supply was pretty well picked over, but we took what was still available: one string of little multi-colored lights, one string of four white snowflakes, and one string of miniature midnight-blue bulbs.

Milt put them up where I decided each should go—the little

multi-colored string woven through the top of the wrought iron fence in front of my duplex to blink back the darkness; the strand of snowflakes fastened to the awning over the window of the guest room that had become his study; the blue set framing the dining room window to bless our dinners that were now being eaten in the earlier evening darkness.

I was mesmerized as each came on at dusk, their tiny solar panels fully charged from the bright, if brief, daily sun. Magically, the snowflakes were often still on at dawn—as if inviting a rare snow to our side of the Las Vegas Valley. The mystery of the sun itself held our collective attention during the season of light. I had always embraced the phrase "in the light of truth." Now suddenly it shifted to "in the truth of light."

And what was that truth? That the sun's Light had brought forth all life on our planet, including our own!

I especially loved sitting in the dark with the gentle glow of these colored lights to focus my winter meditations. Knowing that they were powered by the sun itself connected me with our ancient ancestors who watched and worried while the sun's light diminished and the earth grew cold and barren in the deepening seasonal darkness. For if the sun disappeared for good, all life would cease to exist!

No wonder they created rites and rituals to entice the sun's return! In fact, nearly all of the Christmas customs we participate in today reflect human celebrations carried on for eons before the *Sun* became the *Son*: giving gifts, lighting lights, enjoying special foods and music. Then as now people welcomed the winter solstice as the season of peace.

Historical responses to winter's darkness abound in my home, and include a Hindu Diwali star, a Jewish menorah, and a Solstice candle, along with the single strand of Christmas lights I purchased while in seminary to decorate the large leafy plant in the window of my dorm room.

Now it felt absolutely fitting to have special solar lights to commemorate my new/old relationship with Milt. We'd met decades ago at a summer conference. Because we were both networking in peace, we kept showing up in the same workshops and meetings. When we finally spoke to each other, we discovered we were both involved with the grassroots lobbying effort to establish a federally funded peace academy; I was on the board in D.C. while he was active through his church in Denver. Now, twenty-five years later, we were amazed to be able to celebrate the season of peace as a committed couple.

Together we spent the holidays transfixed as our outdoor lights turned themselves on in the descending darkness, miraculously harnessing current sunlight, unlike the electric holiday lights powered by sunshine that was buried during the Carboniferous Age 400 million years ago.

Back then there were huge amounts of carbon dioxide in the atmosphere, possibly because of volcanic activity, which plants devoured in order to grow. They inhaled the CO_2, then captured the energy of the sun to create photosynthesis, which broke the two atoms of oxygen from the carbon and allowed the plant to manufacture the carbohydrates that made up their roots, stems, leaves, fruits, and nuts. Meanwhile, both the energy concentrated by the plants, plus the Oxygen (O_2) they released into the atmosphere, eventually made the evolution of human life possible.

When the massive amounts of plants spawned by high levels of CO_2 died, they were buried deep in the earth, taking their carbon with them. Dug up as coal, natural gas, and oil, then burned for their energy, these became the fossil fuels through which humans began releasing carbon back into the atmosphere several hundred million years later.

In the process, we are reheating the earth and changing the stable climate upon which our own lives depend. But why burn ancient sunshine, when current sunlight is so readily available? Our little solar-powered Christmas lights invited no end of speculation of what else was possible.

It was such an elegant idea: capturing photons of light, converting them into electrical energy, storing the energy up in an itty-bitty battery, then turning the LED bulbs on at dusk, and glowing long into the night. We marveled at the miracle of it, and speculated about solar energy replacing fossil fuels. Milt shared how he had put solar panels on the roof of his home back in 1979, though they were to harness heat not light. He did know that harvesting light was possible because, as an aerospace engineer, he worked on spacecraft that used solar to power its systems while in orbit. But at that point in time, photovoltaic panels were too expensive for home use.

Then he'd found himself sitting in long lines at the local gas stations because the oil-producing nations had deliberately decreased the world's oil supply after yet another war in the Middle East, for the second time in the same decade! Determined to make our nation energy independent, the Carter administration had funded the exploration of alternative sources for fuel, and, for a while, Milt worked with a shale oil exploration company. Solar and wind energy sources were also being encouraged and funded for development. Carter put solar-thermal panels from the White House roof!

Milt was determined to do likewise. That he did not have a south-facing roof didn't faze him. He literally built an addition to his Denver home primarily for the extra roof space! The panels he put up there pulled heat from the sun and dumped it into a rock box he'd constructed in the basement: ten tons of rounded river rocks stored the day's sunshine, then released it into air that was passed through them whenever home-heat was needed. It turned out not to be terribly efficient, although he did save 30% on his heating bills at the time. But most of his electricity still came from burning fossil fuels. Research and development on alternative energy sources essentially ended when oil became available and affordable again. And Reagan removed the solar panels from the White House roof, signaling that the energy crisis was over. Or was it?

Scientists were beginning to warn about the finitude of fossil fuels: once what had been laid down during the Carboniferous Age was dug up and burned, there was no more to be had. And it would take millions of years to create another supply. Humans were going to have to find an alternative energy supply pretty dang quick, geologically speaking.

Plus, our burning of fossilized sunshine was creating another crisis, this one with far more serious implications: by destabilizing the climate and warming the planet we were putting ourselves on the path to eventual extinction. In spite of the fossil fuel industry's campaign of denial and duplicity, ordinary citizens were getting the message that by ceasing to use coal, oil, and even natural gas to power their lives, they stopped perpetuating the problem.

When Milt returned to Denver for the summer after our solstice solar light adventure, he was determined to go "new" solar. Sunshine, after all, was inexhaustible! It fell freely upon everyone. Photovoltaic panels had become affordable and available, and there were several local companies eager to install them.

The contractor he chose filed for permits with the local utility company, a regulated monopoly that had developed specific guidelines for the amount of solar a customer could install: only 120% of the previous year's consumption was allowable. The utility company could make this demand because a home had to stay connected to their grid in order to have power when the sun wasn't shining. When the sun *was* shining, the homeowner went from consumer to prosumer, putting electricity *into* the grid. A net meter would calculate the amount of energy put in and taken out of the utility company's grid. The company would pay or credit Milt for any excess he generated. The additional 20% allowed Milt to replace his natural gas furnace with a ground-source (geothermal) heat pump furnace that was then powered with solar-electric, making his home essentially "off the grid."

Recognizing that "transportation" created around 25% of his carbon footprint, he wanted to add an electric car that would require more solar-electric panels than he qualified for. Unfortunately, six months of that year had been spent at my home in Nevada, so Milt's energy usage was a fraction of what it had been annually for the forty years before that!

One solution was to leave the thermostat set at 70 degrees over the next winter to run up the electricity usage.

But doing so felt like such a waste because there would be no one home. Another solution was to buy the electric car and charge it using power generated by burning coal and natural gas in order to create a history of higher use. This created a disconnect: the point of "going solar" was to limit carbon emissions, but to qualify for more solar would require emitting even more CO_2 for a full year.

Meanwhile, even though each state had begun to mandate that utility companies get a certain percentage of their energy from non-polluting renewable sources such as wind and solar, this usually resulted in a token effort of 2-3%. And instead of responsibly responding to the new challenges of the time…such as switching over to storing renewable energy for when the wind stopped blowing and the sun stopped shining…the utility companies dug in and fought back against a perceived loss of corporate profit caused by customer-generated solar energy.

Would we play their game and sign on to the fight?

We sat at Milt's dining room table, talking it over.

Along the top of three of the dining room walls was spread the entire story of the universe, beginning with the Big Bang. We had put it together as one of the first things we did as a new couple. One wall depicted the birth of our solar system; the next wall showed the evolution of life on Earth. On the third wall, the human narrative unfolded, from walking out of Africa to landing on the moon. The final panels were a sickly green that reflected the pollution we had wrought upon our beautiful blue marble so recently photographed from space. The fourth wall was blank, a page yet to be written upon. What would we do to help determine how that part of the story unfolded?

We were both getting up there in years. WHY would we expend our diminishing life-energy on yet another struggle against the powers that be? Well, why *wouldn't* we! When my third grandchild was born, I'd taped a little notice on my refrigerator that read: start college fund, revise will, undo global warming. Now my fifth grandchild was on the way, as was Milt's first great-grand stepchild. How could we leave behind a polluted planet that couldn't sustain them! Plus, our recent involvement with the Occupy Movement had reawakened our outrage over the greed and maliciousness of our too-big-to-fail corporate systems, and the corrupted government officials they had bought off via campaign contributions.

We *would* take on the fossil fuel industry, through the utility company monopolies in both Colorado and Nevada. Milt would hike up his home energy usage for the next year, even when we weren't there all winter, and qualify for ten more solar panels the following summer to run an electric car.

If you don't use natural gas, there is no need to destroy whole neighborhoods along the Front Range to frack for it. If you are not using gasoline, there is no need for a transcontinental pipeline to carry tar sands oil to refineries. For me the bug-a-boo was coal. I'd spent time in Charleston, West-by-God-Virginia, when my former husband briefly lived and worked there. While there I attended a daylong seminar on death, dying, and grief. As a newly-minted minister, I appreciated anything that would augment my own work. But what I learned there floored me: grief was endemic in that state because their mountains were being decimated for coal. Whole tops were being blasted off defining landmarks, and anyone who ventured to the sites to have a closer look was stunned speechless by deep grief over their loss. Yet West Virginia was coal country; coal was the basis of their economy, and Coal was King. So why is it the poorest state in the nation? Hint: the profits from the state's major resource have funded New York museums, universities, and libraries that were philanthropically endowed by those who owned the coal mines in West Virginia but lived elsewhere. The citizens of West Virginia were left with decimated mountains, coal rubble polluting their rivers, people living in poverty in their hollows, and coal miners suffering from black lung disease who had to fight with the coal companies for their guaranteed worker's compensation.

I live in Las Vegas now, where a main source of electricity comes from the coal-fired power plant 70 miles north. While part of Occupy Las Vegas, we'd driven up to the area: the plant sat ominously beside the Moapa Paiute community. No wonder the Paiute elders had led the march to the utility company's headquarters when we demonstrated for and demanded clean energy: Earth Justice had put out a film called *An Ill Wind* that showed the effects of coal ash on the people downwind from the plant.[26] Especially disturbing was the sight of children suffering from asthma and using inhalers in order to breathe. The elders, after breathing in toxic coal ash for fifty years, were suffering and dying from heart and lung diseases. Because the population was too small to warrant an official government health study, officials continued to ignore the people and the problem. On Bad Air Days everyone was supposed to stay inside while the wind blew the coal ash over everything. How could this ethically be expected of a people who revered the earth and had a spiritual need to be outside in the sunlight and air, communing with land and water?

The Sierra Club, Occupy Las Vegas, Earth Justice, faith groups, and concerned Vegas citizens joined in the effort to shut down the coal-fired plant. Plus, the Moapa Paiutes were getting ready to install arrays of solar panels on their land, so they could replace the *lost* power.

In preparation for installation, the tribe was relocating the endangered desert tortoises as they came out of brumation that spring when we all trekked out to the site to celebrate Earth Day. We held paper solar panels over our heads to simulate what was soon to be. The Sierra Club had provided them because closing the Reid Gardner power plant was a major part of their Beyond Coal Campaign.

Putting solar panels on my own rooftop got complicated. I belong to two Home Owners Associations: one for the entire Sun City Community, the other one because I live in a duplex. If you were to drop the main HOA's loose-leaf binder of rules and regulations on your foot, you'd need a walker for sure. Drop the secondary HOA's binder of community covenants and regulations on the other foot and hopefully you'd have access to a wheelchair. While you sat there reading through both tomes, you would not come across one word about putting solar panels on your rooftop.

How could that be, especially in southern Nevada, where *sunshine* is our major natural resource? As one state lawmaker puts it, "There's more sun in Nevada than heat in hell!" In fact, the state legislature had recently decided that a homeowner's decision to go solar trumped any and all Home Owner Association regulations.

I found a solar installer, assembled my electricity bills so they could size my system for what the utility company would allow, and, as they pulled permits, I sought formal permission from my two HOAs. They couldn't say no, but they could require me to agree to remove the panels when I moved if the new owner didn't want them, repair the roof, and turn over my last-born grandchild…well, okay, maybe not that one. But going solar was not simple.

I jumped through the hoops because the image of a Paiute child breathing through an inhaler haunted me whenever I flipped on my light switch. Pushing to put up 8 solar panels for when Milt got his electric car and needed to plug in over the winters spent in Nevada, I wasn't surprised that my local utility company said no...5 were all I qualified for, and thus allowed. It felt like I was being punished for my ultra-careful use of their electricity—I used minimal air conditioning, loved sitting in the dark instead of having lights on everywhere, etc. So of course, the powers that be could and would undersize the system I was allowed.

When I did finally get my panels in place, the inspector who had to sign off on them kept the installer sitting idle in my garage for 10 hours waiting for him to show up. Then the utility company took its sweet time (five months!) to install the required net meter, claiming they were swamped with requests. I called it passive-aggressive behavior.

At issue was energy control. Rooftop solar is the ultimate act of democracy: for self-determination, for energy efficiency, and for good economic sense. Small-scale solar gardens have begun springing up in vacant lots to provide clean energy to apartment dwellers and low-income residents in Colorado. We're working to make that happen in Nevada too.

While the major energy provider in southern Nevada was ready to close its coal-fired power plants, putting up large solar arrays seemed the only way to retain control over energy production, delivery, and thus profit. The problem with industrial-sized solar splayed across the open desert is the disruption of the fragile eco-system that is utterly unnecessary if rooftops are utilized for producing energy rather than wasting heat.

Once my system was up and operating, I could watch it through my computer. Not only could I see it wake up and start producing as the sun rose in the sky, but I could also see how much CO_2 I had NOT dumped into the atmosphere! By Milt's calculations based on research he found on government websites, going solar dropped my carbon footprint by 70%!

But for me it was more like creating a handprint. By gathering up current sunlight, sharing with my immediate neighbors what I wasn't using, then sending the rest to the grid to be withdrawn when needed after sunset, I was doing something proactive for the common good and on behalf of the climate, as well as the Paiute children and all of our children.

Plus, greeting the sun both visually and virtually became part of my morning spiritual ritual. I've always lit candles to the dawn as part of my morning meditation, and now as a modern human watching the rising green curve on my computer screen, I joined other humans across all time who had paused to pay homage and attention to the Sun.

From adoring Pharaohs in ancient Egypt to chanting priests in the temples of Peru, Andean flute players in Mexico, golden bell ringers in China, and drummers in the Congo, we humans have always known that we depend on the Sun for the source of our life. Long before electricity lit up the night (from my window I can see the neon lights of the Vegas Strip on all night and so bright they can be seen from out in space) the Sun mattered! Years ago, I had the privilege of working with a Hopi elder to set down in writing their foundational custom of presenting a new child to the Sun twenty days after its birth. Into the soft spot on the infant's head, the Sun would pour the "wisdom of responsibility, tolerance, respect, contribution, and sharing." Then the baby's own life plan would be poured into the hole before it sealed closed. The child thus belongs to the Sun.

Greeting the Sun each new morning, Hopis tap into its wisdom and seek its guidance day by day. And from the Indians on the banks of the Ganges come the Sanskrit hymn to morning I have long loved:

Look to this day! For it is life, the very life of Life. In its brief course are all the verities and realities of your existence; the bliss of growth, the glory of action, the splendor of beauty; for yesterday is but a dream, and tomorrow is only a vision; but today well lived makes every yesterday a dream of happiness, and every tomorrow a vision of hope. Look well therefore to this day. Such is the Salutation of the Dawn.[27]

Through my awakening solar panels that not only honored the Sun by greeting it and gathering up its energy, I got to save and savor the amazement of the sun's light as I tallied up the number of kilowatt hours being produced for human use. Three years after installing solar, the read-out from the panels being monitored on my computer by the Sunpower company showed that so far my "lifetime environmental savings" included:

CO_2 Emissions avoided: 8.7 tons
Miles not driven: 18,690 miles
Gasoline not used: 883 gallons
Coal not burned: 8,432 pounds
Crude oil not used: 18 barrels
Mature trees grown: 201 trees
Garbage recycled: 6,203 pounds

We humans are primarily a tool-making species, and solar panels are merely a modern tool to harvest current sunlight without causing climate-changing consequences. For burning coal to release its ancient sunshine is a one-time act that leaves a toxic mess behind, and poisons air, water, and soil. Solar panels can cleanly provide electricity for many years, then be recycled in solar-powered factories, reflecting the emerging new paradigm of "from cradle to cradle."

Yet the utility company fended off rooftop solar any and every way it could, beginning with a classic bait and switch through which it removed or reversed initial incentives and net metering credits and reductions in solar customers' energy bills based on how much energy they put into the grid, all while claiming that solar prosumers were not paying their fair share of infrastructure costs that were already under contract. Never mind that upgrading infrastructure for more fossil fuel energy sources became unnecessary when replaced with solar generation!

If corporations were indeed people as the Supreme Court decided, then the utility companies were not being good citizens: they wanted rights without responsibilities. We the solar-prosuming public pushed back! State laws and incentives had opened up a booming solar industry, with several companies offering a power purchase agreement plan, which essentially meant they leased your roof space for solar panels and gave you a break on your electricity bill. Most of southern Nevada's nearly 15,000 solar customers were on that plan. I was among the minority who had purchased the panels outright, and taken the tax break myself.

Now we the people all crowded into the Public Utilities Commission hearings on rate increases; those of us unable to get inside demonstrated out in the parking lot. We wrote letters to the PUCN, the governor, our legislators, and the newspapers. Unable to hold back the tide of rooftop solar, the utility company resorted to pitching neighbor against neighbor, claiming that solar customers were harming regular customers by causing their rates to increase. What? How did putting clean energy into the grid harm rather than help? The fallacy with this piece of propaganda was this: we *were* paying service fees to stay connected to the grid. But not enough, the utility company insisted: they needed to increase fees to pay for future fossil fuel burning infrastructure.

The PUCN honored the P in its title and scheduled more public hearings. Being accused of harming rather than helping my neighbors pushed me over the edge. I would have to testify in person before the PUCN. Determined to keep my anxiety isolated in the pit of my stomach and out of my voice, I sat before the intimidating utilities commission and introduced myself as a solar prosumer as well as a retired minister.

Then I explained that for me, going solar was never about money; it was about morality, and I found being accused of harming people to be insulting and disingenuous…perhaps even a projection by the energy company for the harm they did to people and planet in the name of profit and for power. We prosumers were meanwhile offering the utility company a gentle way to comply with the Clean Power Plan that had been put in place as part of our country's commitment to reducing emissions we'd agreed to at the most recent U.N.

climate summit. Climate change was the bigger issue here, with the future of the planet at stake. We humans had been clever enough to create the problem, surely we were wise enough to fix it!

That testimony was a month before the 2016 election. Then, as the United States, a main perpetrator of the climate problem, retreated from taking responsibility to help solve it, Nevada plodded forward.

Our state legislature passed eleven clean energy bills in 2017. The governor signed all but two of them into law, and the solar companies that had left Las Vegas, taking 6000 jobs with them, began to return. My phone rang off the hook all day long with solar vendors trying to save me money on my electricity bills. These days I can drive almost anywhere in the valley and spot solar panels on more and more rooftops. And in the northern part of the state, Tesla is manufacturing power walls to store sunlight in battery packs for people's garages.

Other southwestern states have been watching Nevada's saga. In fact, when filmmaker Jamie Redford travelled across the country exploring how clean energy options were faring, he put Nevada on his schedule.

Trusting that when people recognize there's a problem their initial response is "what can I do to help fix it," Redford had embarked "on a personal journey into the dawn of the clean energy era as it creates jobs, turns profits, and makes communities stronger and healthier."[28]

From Georgetown, Texas to Buffalo, New York, Redford showed how unlikely entrepreneurs were developing clean energy solutions, and explored issues of human resilience and social justice in a way that provided hope for the future. The final third of his film *Happening, A Clean* Energy *Revolution*, was devoted to Nevada's solar fight.

Watching it revealed something I didn't know: the federal government too had been watching what was happening in Nevada, apparently to gage the public's interest and commitment to the clean energy revolution that had to happen, sooner rather than later.

Nevada had even put forth a ballot initiative to increase the renewable portfolio standard required by the state to 50% by 2030. It passed by 60%. Clean, renewable energy was becoming a reality!

And outside my back-patio door, a solar-powered clear plastic angel I've stuck in a flowerpot blinks blue, green, and red from dusk to dawn all year long. I wink back in salute to the magic of the Sun.

FOUR. GREEN
BRIDGING

"The dinosaurs never saw that asteroid coming.
What's our excuse?"—Neil DeGrasse Tyson

"WHAT IF WE CALL THIS ONE 'BIRGITTE?'"

Unable to answer me because he was blowing his life-breath into the mass of green plastic, Milt simply raised his eyebrows.

"You know...after the kick-butt female Prime Minister on *Borgen*"...which meant "government" in the Danish T.V. series with English subtitles we'd been following for three years.

Any answer got lost in the process of bringing to fullness the replica brachiosaurus: chunky body, stocky legs with plodding feet, long slender tail, and, finally, the enormous neck raising the tiny head into our winter session living space.

Catching his breath, Milt finally responded, "I'd thought of her as being Bridget 2, but I guess Birgitte would work too."

It should work just fine, because that was the Scandinavian form of Ireland's Bridget, the Celtic goddess of strength. We'd named our first dinosaur Bridget because the name seemed to echo "bridge," and we saw her as the bridge between the last mass extinction and the coming one we humans seem hell-bent on causing. This insight had come to us after watching a new version of Carl Sagan's *Cosmos*, in which we had heard astrophysicist Neil deGrasse Tyson proclaim:

> We just can't seem to stop burning up all those buried trees from way back in the carboniferous age, in the form of coal, and the remains of ancient plankton, in the form of oil and gas. If we could, we'd be home free climate-wise. Instead, we're dumping carbon dioxide into the atmosphere at a rate the Earth hasn't seen since the great climate catastrophes of the past, the ones that led to mass extinctions. We just can't seem to break our addiction to the kinds of fuel that will bring back a climate last seen by the dinosaurs, a climate that will drown our coastal cities and wreak havoc on the environment and our ability to feed ourselves. All the while, the glorious sun pours immaculate free energy down upon us, more than we will ever need. Why can't we summon the ingenuity and courage of the generations that came before us? The dinosaurs never saw that asteroid coming. What's our excuse?[29]

It's getting ever harder to deny that humans are causing the planet to warm to temperatures not seen since the time of the dinosaurs. Extreme weather has already been wreaking havoc, and exacerbating the wildfires along Colorado's front range. In fact, we'd been trying to get used to choking on smoke during our summers at Milt's home in Denver. But we were also addressing the underlying problem through a plethora of activist groups there.

In fact, we'd bought the first Bridget in the dinosaur museum gift shop in Morrison in time for a tabling event with Citizens' Climate Lobby. Organized around the purpose of creating political will for a livable future by moving local activism to national action, this citizen-lobbying group is pushing Congress to pass a carbon fee and dividend plan. This would make fossil fuel companies pay for the carbon they emit and offset the costs of the damages done to public health and planetary wellbeing.

Currently, fossil fuels receive millions of dollars a day, billions of dollars a month, trillions of dollars a year in subsidies from local, state, and national governments and through international agreements. Such subsidies include actions that lower the cost of fossil fuel production, raise the price received for the product, and artificially lower the price paid by consumers via direct funding and tax giveaways. Loans and guarantees at favorable rates, price controls, government provided resources like land and water (such as leases on our public lands) at below-market rates, plus research and development. None of these subsidies carries any expectation that the fossil fuel industry take responsibility for the "unintended consequences" of their product.[30] Thus the public must absorb the health and environmental costs of the industry doing business. Often this includes cleanup when a company goes bankrupt and walks away from the toxic mess it has created.

Levying a fee on the greenhouse gasses they dump into the common atmosphere would acknowledge the *externalized* price of their product and help level the playing field for the needed shift to renewable energy.

Plus, each household would receive an automatic dividend check divided up from the collected fee to cover the increased product price that would be passed on to consumers at the pump. If you drove an SUV and put $100 worth of gasoline in your tank, you'd get back exactly the same amount as someone who drove a Prius and filled the gas tank for $10. With the true cost of fossil fuels on the line, we dared to hope that more of it would be left in the ground, and thus kept out of the atmosphere.

We got involved and helped set up the Citizens Climate Lobby exhibit table that would be part of the Juneteenth celebration held annually in a predominantly African American part of Denver. Originally, it had commemorated the delayed arrival of the news of the Emancipation

Proclamation. Now its focus was on education and empowerment, and we'd printed out small pictures of Neil deGrass Tyson with his words about climate change. On each we affixed a small plastic dinosaur, to underscore his point. Our large inflated green brachiosaurus was strapped to the top of the shade tarp, to catch children's attention.

After that, Bridget became a significant, if silly, feature in the Denver dining area, where we had murals of the entire 14-billion-year story of the universe high on one wall, the 4.6 billion year earth story on the next, and the human story along the third. She fit right in, and made herself at home. We became known as the "dinosaur couple" because we handed out little plastic ones at climate-action events we took part in.

For the two months that Milt and I Skyped every evening while I was back at my home in Nevada and he was still at his in Colorado, Bridget presided over our conversations. Naturally I asked him to bring her along with him when he came to Vegas. Instead, he left her in Denver, and purchased a sister dinosaur for the Nevada space. Our pet brachiosaurus had become another bridge, the one between the two places where we engaged in climate activism!

Bridget/Birgitte was also the spitting image of the dinosaur seen at gas stations owned by a western company that recently celebrated one hundred years of production and profit, with plans to expand operation coast to coast, when they should be leaving their product in the ground.

Meanwhile, our *symbol* of the extractive industry's pillaging, poisoning and plundering of planet and people for power and profit was banished to the garage, to off-gas her eye-watering, throat-choking petroleum by-product smell. The irony struck me every time I opened the door to the garage: the mass of green plastic made from the very oil that geologists were searching for beneath the bones of these beasts that had died hundreds of millions of years after the oil field was formed during the Carboniferous Age. If she were life-sized, her tail would reach into the neighbor's garage across the street, and her long neck would raise her little head right up through my garage roof where solar panels sit gathering current sunlight. We are not stuck with fossilized energy!

Wind and solar can replace coal-provided electricity, and geothermal earth energy can be exchanged for heating and cooling and replace the hot water heater and furnace behind her. But that will take human will. We have to want to cross the bridge between scientific reality and the fantasy world that burning fossil fuels has enabled. Could this ridiculous green dinosaur help carry us across the great ideological divide?

Nothing has stirred the human imagination as much as finding the remains of these massive creatures: they have been the stuff of myth and legend for centuries. Historically,

President Thomas Jefferson had secretly hoped that the explorers he sent across this uncharted continent would come across dinosaurs still roaming the plains.

In a way, they were. As the landscape grew drier and shed its vegetation, huge bones stuck out of naked rocks. What did the pioneers make of them? How did indigenous peoples explain their presence? Who decided that these ancient beasts were green? They could have been as pink as the one outside a motel in Vernal, Utah. And why not: the high desert gives back the shades of sunrise and sunset as a warm coral-salmon-pinkish glow, as much from within as from sky. Why couldn't dinosaurs have been that shade of pink, or even multi-colored?

We spent an overnight in Vernal while visiting Dinosaur National Monument that straddles the Colorado Utah border. This area has been ground zero for oil and gas extraction for decades. Ever since the national oil crisis in the mid-1970s sparked the exploration for alternative energy sources, energy companies have been exploiting the fossil fuel reserves trapped in shale rock under western Colorado and eastern Utah. They do this mainly by hydraulic fracturing, a process by which massive amounts of water are mixed with sand, kept soluble by a cocktail of proprietary chemicals, then injected by force into ancient rock in order to break and hold it open so that it can release any residue of oil or trapped natural gas.

Let's be clear about one thing up front: natural gas is *not* the *bridge fuel* to renewable energy that has been promulgated by our politicians and the industry alike. This is a popular myth we desperately want to believe so we don't have to change our life ways: surely, we can simply safely substitute natural gas to heat our homes and run our city buses the "cleaner, greener way." Climate problem solved!

Alas, yes, gas burns cleaner than coal for generating the electricity for our massive electricity grids. But its production releases methane (CH_4), a more potent greenhouse gas than CO_2, and makes the climate problem *way* worse instead of a bit better. The other dangerous effect of fracking first burst into public awareness in 2010 with the Josh Fox film *Gasland*—who can forget watching a homeowner light a match under his faucet and having the water burst into flames?[31]

Fox began this project when he himself received a letter from a gas company offering to lease his family's land in Pennsylvania for $100,000 to drill for natural gas. This prompted him to research how communities across the country were being affected by this industry that was largely exempt from oversight and regulation.

Spending time with citizens in their homes and on their land in Colorado, Wyoming, Utah, and Texas, Fox listened to their stories of how natural gas fracking was affecting them, from a variety of chronic health problems to contamination of their air and water wells and surface water.

Hydraulic fracturing had been exempted from the Clean Water and Clean Air and Safe Drinking Water Acts in the Energy Policy Act of 2005, pushed by then Vice President Cheney, formerly chief executive of a major fracking company. Now shiny white fracking company trucks salt the landscape around the Dinosaur National Monument. After enjoying a few days there paying our respects to our *terrible lizard* cousins the summer of 2014, Milt and I stopped for lunch in a diner. There was a fracking company's van in the parking lot. Inside, I couldn't help but notice that the worker wearing the fracking company shirt was gingerly eating his burger with a knife and folk. What was he afraid of?

Call me suspicious, but then again, western Colorado resident Theo Colburn, an environmental health analyst best known for her studies on the health effects on endocrine disruptive chemicals, had first talked about the public health issues of gas development in 2005. Back then she emphasized the need for full disclosure of the chemicals used to produce and deliver natural gas. In the last year of her life (2014), she published an air quality study near natural gas operations and developed a continuing medical education course with recommendations to the health care community on air emission exposure. [32]

As ordinary people caught on and pushed back against this health danger by refusing to lease their land, they discovered that the gas companies could take it for free via eminent domain. Concerned citizens across the country began protesting and resisting fracking wells and compressor stations and invasive pipelines carrying natural gas through their neighborhood ecosystems. For instance, an activist-grandmother became relentless in trying to expose the risks associated with the fracking boom that had burst upon Pennsylvania...her Susquehanna Country sat on a shale formation that was one of the most active areas for the natural gas rush.

She visited frack sites, called in health and environmental regulators at perceived violations, and organized bus tours of the sites for anyone interested, from celebrities (Yoko Ono, Susan Sarandon) to visiting Canadian officials. None of that was illegal or presented a public danger. Even by the fracking company's own admission, she seemed to be more of a nuisance than an actual danger. Still, the industry secured a court injunction barring her from all property owned or leased by the company. The ban effectively made off-limits the new county hospital, the restaurant where she takes her grandchildren, the supermarket and drug stores where she shops, the animal shelter where she adopted her dog, plus the bowling alley, recycling center, golf club, and lake shore.

"They might as well have put an ankle bracelet on me with a GPS that tracks wherever I go," she told reporters, "I feel like I am some kind of a prisoner, that my rights have been curtailed and restricted." The ban represents one of the most extreme measures taken by the oil and gas industry against protesters operating peacefully and within the law. [33]

Meanwhile, along the Front Range of the Rockies, where dinosaurs once roamed and now densely populated cities sprawl across the top of an ancient basin containing massive amounts of oil and gas, fracking sites have popped up like poisonous mushrooms beside homes, schools, playgrounds, and public spaces. When citizens expressed concern over the toxicity of the fracking fluid, the governor, an oil and gas man himself, drank from a jar of the stuff presented to him by a fracking company official who promised it contained only food-grade additives. Claiming he did it to earn the trust of the industry so that he could regulate methane levels, he now boasts that his state has the most comprehensive hydraulic fracturing rules in the country because companies are required to post the ingredients of the mixture—yet are also allowed to protect their intellectual property by claiming that the chemicals were trade secrets. State-mandated setbacks have been raised from 150 feet in rural areas and 350 feet in urban settings to at least 500 feet from homes and other buildings statewide by 2011.[34]

By then I'd begun spending summers in Denver. In 2012 I became friends with a concerned, committed woman who'd been worried about species extinction since watching a documentary on endangered polar bears. As the polar bear mother struggled to keep her twin cubs safe on the melting ice, my friend looked over to see tears running down the face of one of her own twin sons.

"What are we going to do?" he demanded to know.

Her answer included taking Al Gore's *Climate Reality Project* training and giving presentations on the scientific causes and effects of climate change. And because fracking was the local contributor to the climate crisis, she started showing up at public meetings held by the Colorado Oil and Gas Conservation Commission as it considered new permits.

At one such meeting, youth from Boulder, still too young to vote, tried to reason with the commission before it rubberstamped more permits. When the commission refused to listen, the young people commandeered the dais and voiced their concerns anyway. Once the youth left, the commission went ahead and approved the fracking permits. Outraged, my friend was moved to form a group she named *Wall of Women* that would literally stand behind the state's young people, letting them know there were adults who cared about their future and had their backs in the fracking battle.

Carrying her *Wall of Woman* banner, she's become a persistent presence at fracking demonstrations and protests along the Front Range, most recently at the Bella Romero Academy in Greely, when the industry planned to locate a multidirectional wellhead right beside the school playground. By then it was 2018 and the politically determined state laws allowed drilling sites to be located 500 feet from homes and 1,000 feet from schools. Meanwhile, school boards had voted to require 2,000-foot setbacks, but the oil and gas companies refused to honor them.

This became a glaring environmental justice issue because Bella Romero is 89% Hispanic, and 92% of the students are also from low-income families. Plus, the community surrounding the school is within the 90-95% range nationally for linguistically isolated communities. The proposed well pitted the oil and gas company against a disadvantaged community in a David vs. Goliath battle that drew national attention.[35]

The greater Denver community mobilized against putting yet another sacrifice zone on the altar of the corporate economic model. Armed with scientific studies revealing increased cancer, asthma, and nose-bleeds within a 2,500 foot range, and frightened by fifteen gas explosions and fires within a year, plus outraged that three schools were doing emergency evacuation drills, mobilized and determined Colorado citizens managed to get a 2500 foot setback initiative on the state's midterm ballot. The industry panicked and responded by claiming that 2500-foot setback requirements would end the oil and gas boom and cost the state millions in dollars and lost jobs.

Truth be told, horizontal drilling techniques can reach the same pools of gas by using longer pipes, and the industry can still have access to most of the state's oil and gas reserves. In reality, the mining industry accounts for only 1% of the state's tax revenue and job base, as more and more people were moving to Colorado for quality of life values that include outdoor recreational opportunities, a whole industry in itself.

Unlike the boom-bust cycles of the early days of the state, when settlements grew up around mines that became ghost towns when the mines played out, today's population centers grew up around other economic opportunities. Then the extractive industry found them.

Basically, the industry seemed afraid that to lose the vote would be conceding that there are the public health risks associated with fracking. Then they might be held accountable and legally responsible for the inevitable consequences. "Fracking is the worst thing I've ever seen," claims Dr. Sandra Steingraber, a biologist who has worked as a public health advocate on issues like breast cancer and toxic incinerators. "Those of us in the public health sector started to realize years ago that there were potential risks. But then the industry rolled out faster than we could do our science."[36]

In recent years, as the practice expanded from rural lands to backyards, farms, and within sight of schools and sources of drinking water, "now we see those risks have turned into human harms and people are getting sick. And we in this field have a moral imperative to raise the alarm," Steingraber insists.

In her keynote address to the 2018 Permanent Peoples' Tribunal on Human Rights and Fracking,[37] she warned that science alone does not change policy; activism must be added to make change happen. It's no wonder she joined my friend and others at a protest outside the Bella Romero Academy! When pressed to comment on the Colorado governor's frack-fluid-cocktail stunt, she shook her head and insisted that, had he drunk the actual stuff, he would *not* be walking around still bragging about it today.

To protect its 31-billion-dollar business, the oil and gas industry spent over a million dollars to get enough signatures for its own ballot initiative: a constitutional amendment that could give oil-and mineral-rights owners the ability to sue and collect millions if government action diminished the value of those rights. In short, they were trying to protect their profits if their assets had to remain in the ground.

Meanwhile, they accused the activists of having a hidden agenda: to kill the oil and gas industry altogether. If that was *not* the "fractivist's" point, perhaps it should have been...if there is to be any hope for a habitable planet in the near future. Instead, the federal government has moved to exempt fracking from even minimal regulations on the methane that's 80 times more potent than CO_2 as a heat trapping greenhouse gas.

Trying to understand how profit can continue to trump human and planetary well-being, I find it helpful and hopeful to picture the dinosaurs' reptilian brain layers, that primitive consciousness that automatically responded to outside stimuli by fight, flight, or freeze. Even if they had seen that asteroid coming, dinosaurs were ill equipped to do anything about it except run, hide, or ignore their new reality. They were trapped, not just by the event itself, but by their limited capacity to deal with it.

I'm tempted to make connections between reptilian responses and the corporate mindset obsessing on threats to their bottom line, but that's an unfair comparison. While we know that reptiles are pre-mammalian, i.e. devoid of emotional connectors, new research on dinosaurs suggests that some late members of this long-lasting species (millions of years on earth) tended rather than abandoned their eggs, and kept the young with them in herds. Was evolution trying out the next layer of brain development, the one that burst onto the planetary scene fully with mammals, then us?

For instance, when my youngest grandchild first met Bridget at the Denver house, he was five years old. The awkward mass of green plastic fired Liam's imagination and sparked his compassion so that in no time at all he had placed her in a large wicker basket that he could pull or push from room to room, wherever he wanted to be, so that she would be with him. We knew his bonding was complete when he shared his security blanket, gently covering as much of his large inanimate playmate as possible with the small square of faded blue fabric. I felt a twinge of hope; human empathy and imagination may yet save us from going the way of the dinosaurs!

FIVE. BLUE

SHADOW

"You cultivate the flower in yourself,
so that I will be beautiful.
I transform the garbage in myself,
so that you will not have to suffer."
—Thich Nhat Hanh

ANCHORED TO THE PRESENT MOMENT by the line of
mountains chalked against the horizon west of Denver, I
venture back into the past when I first saw, first hand, the
wrought iron Arbeit Macht Frei (work sets you free) sign
at the entrance to Dachau concentration camp outside of
Munich.

It stopped me cold in my tracks. Would I go any further?
Why should I? Yes, I was working on my M.A. in Peace
Studies while my husband served with NATO forces in
West Germany. Yes, I was trying to understand how such
a civilized country could tip over into its darkest side, and
become a scourge on its own citizens as well as on those of
so many other countries.

Where better to confront that haunting mystery than
at Dachau? I hesitated, then looked to the landscape for

strength. If the weather phenomenon known as the *foehn*, the hot dry wind from the Sahara Desert, really did make the Alps appear to move close to Munich, then I might be able to glimpse mountains from this nearby town. Instead, all I saw was the mostly empty parking lot, where the car license plates were mainly from neighboring countries. Very few in-country locals seemed compelled to visit this historical place back in the late 1970s, except for a handful of grandparents, their young grandchildren in tow.

Taking a deep breath, I passed through the gate and into the horror of man's inhumanity to man. While exploring the psychological aspects of war and peace for my degree, I'd immersed myself in the works of Carl Jung, and was blessed to study with one of his original pupils. Among other insights, Jung claimed that we are all secretly fascinated by the Holocaust because each of us is capable of participating in such cruelty. Most of us deny that dark part of ourselves because "projection is one of the commonest psychic phenomena…Everything that is unconscious in ourselves we discover in our neighbor, and we treat him accordingly."[38]

I tried to keep that frame in mind as I proceeded towards the introductory exhibits. All these years later what I mostly remember is that I never made it through the black and white larger than life photo gallery that began with the burning of books. I didn't need to see it through to the end: the gassing then burning of people.

Fleeing the heart-breaking, gut-wrenching, mind-blowing exhibit, I collapsed on a seat in the courtyard, catching my breath and shuddering at the nightmares forming in my psyche. From where I sat, I could see the buildings that housed the barracks, the showers, and the ovens. Above them all, the Bavarian blue sky was no longer smeared with airborne ash.

Instead, the ground beneath my feet was composed of ash-grey stones. Looking closely, I could make out the soft shapes of tiny bone-grey flowers bravely poking up from the gravel. As much as I wanted to embrace this sign of hope, despair hovered over every molecule of memory still haunting this space. I could feel the presence of a clergy member of my own faith tradition, arrested for preaching against the Nazis from his pulpit in Prague, who underwent medical experiments here. And while we historically think of millions of Jews as the major victims of the Holocaust, in reality, millions of others perished as well: homosexuals, the mentally ill, the mentally defective, gypsies, Russian soldiers, and resisters from many countries. So many Catholics perished here, especially dissident priests, that committed Carmelite nuns now care-take the cemetery grounds.

Without the insights of Jung, I would have found it impossible to understand the inhumanity that we humans are capable of. As he describes it, the repressed part of our psyches that we reject and then project onto others stays buried deep in both the personal and collective unconscious until awakened by something, or someone...like a Hitler.

This paranoid, insecure, delusional little man wasn't taken seriously when he was legitimately elected to the German government at the time when the country's economy had been decimated by the demand for reparations after World War One. Germany claimed it wasn't responsible for that war, and resentment festered as the populace suffered from the resulting humiliation along with the hunger caused by deprivations.

With a toehold on national power and legitimacy, Hitler became a mastermind at triggering the ancient Teutonic archetypes deep within the German psyche, thus unleashing Wotan, the war god of storm and frenzy, upon the world. Mass fear and hysteria became the hallmarks of The Third Reich as it capitalized on the country's collective sense of inferiority and guilt. Scapegoating, persecuting, conquering and purification became national priorities that were perpetrated by a demagogue acting like a demigod, until Germany viciously turned against its European neighbors.

Because the shadow resides in the part of the psyche that doesn't get exposed to the light of day, all the insecurities that mushroom in its darkness take on a life of their own and become autonomous. Consciously unaware of this dynamic, we become powerless over it. Our inner demons become devils, and dinosaurs become dragons as we humans engage in the classic battle between good and evil, the forces of light and darkness playing out in ourselves as well as out in the world. We seem to be possessed, as if the demon/dragon within is holding us hostage in the dank dark cave of human evolution become devolution.

Today, the delusion that burning fossil fuels isn't changing the climate has become the demon/dragon trapping us in the cave of inaction. The window of time to do anything is closing fast, with each extreme climate-exacerbated weather event causing more deadly natural and national disasters. How can we fight these forces that would destroy us?

The human unconscious has been likened to the ocean, that natural element that makes our planet look like a tiny blue marble out in the vastness of space. Much of what we know about ourselves is hidden in our depths, as out of sight and out of mind as the garbage we currently dump into the seas. Point of fact, the unwanted garbage of our lives is as dangerous to our long-term well-being as is polluting the oceans with plastic, toxic chemicals, spilled oil from offshore drilling, and the sinking of atmospheric carbon. Sooner or later we shall have to deal with what we'd rather ignore in ourselves and then blame others for manifesting, on both personal and collective levels.

Recognizing this begs the questions: what is hidden in the collective American unconscious? What is crouching in our shadow that seems to have been newly released by letting loose a bully in the china-shop of our democracy? What demons have broken through to bedevil us?

For a clue, I look to news clips of the current White House occupant sitting in the oval office signing executive orders, most of which reflect the wish list of the fossil fuel industry, a huge campaign funder and profiteer from the greatest crime ever perpetrated against humanity in the whole human narrative.

Over his shoulder hangs a portrait of his professed hero, President Andrew Jackson. Jackson's 1830 Indian Removal Act led to the infamous Trail of Tears for the Cherokee, the Choctaw and the Chickasaw peoples. It also resulted in the Long Walk for the Navajo and Apache, and forcibly re-located a dozen other tribes, causing the deaths of tens of thousands of indigenous people.

When gold was discovered in the hills of North Carolina and Georgia, the Cherokee people who occupied the land simply had to be removed. Even though the U.S. Supreme Court ruled that plan unconstitutional, nothing stopped President Jackson from enacting it anyway. In the Cherokees' version of that travesty as re-enacted every summer on their sliver of reclaimed homeland in the Smokey Mountains, one of their own had saved Andrew Jackson's life during the Battle of New Orleans…an act they've collectively lamented ever since.

Sitting in that audience, I heard the people around me express shock and regret and remorse for what was done in our name so long ago. Having lived so much in the west, I longed to inform anyone who would listen that the same exploitation is happening in our own era, with real estate developers, oil, gas, and coal corporations, uranium and gold mining companies all still grabbing at the remaining native lands.

We would rather stick with the whitewashed national narrative of idealistic intentions that we commemorate on patriotic holidays. One glaring example is the Pilgrim story we lift up each year at Thanksgiving that emphasizes the virtues of thanking god for the harvest that sustained our brave forebears seeking religious freedom. This gesture has come to affirm our sense of being especially chosen to discover the new continent. This was officially known as The Doctrine of Discovery.

In reality, much of the discovered new world was colonized for raw resources, the timber, sand (for making glass), indigo, etc. to be sent back to a mother country that had badly overshot the carrying capacity of its own limited ecosystems. Discovery meant dominance.

Truth be told, even the deeply religious Pilgrims had no problem helping themselves to the caches of corn that the native peoples had laid in store for the harsh New England winter. Manifest Destiny soon became the justification for helping ourselves to the rest of the continent as well, with the help of the U.S. Cavalry. We called this "taming the west."

"Concentration camps were not invented in Germany," Hitler claimed in 1941. "It was the English who were their inventors, using this institution to gradually break the backs of other nations." Colonialism perpetrated the domination mindset that then made indigenous peoples into less-than-humans to justify their demonization and destruction.[39]

Hitler noted early on that the white settlers in America "gunned down the millions of redskins to a few thousand." As early as 1921 he talked of confining Jews to concentration camps not unlike the American solution of Indian Reservations. For inspiration and confirmation, he had only to look at the presidential actions of Abraham Lincoln.[40]

In 1862, Lincoln ordered the largest mass execution in our history with the hanging of 38 Indian men after the Dakota War. Next, he nullified all the Treaties with the Dakota people in Minnesota. Then he granted himself authority, without treaty or negotiation, to remove the tribes from that state. Finally, in 1862, he signed the Homestead Act, then the Pacific Railway Act. The Homestead Act provided 160 acres for any American citizen willing to homestead land in the west for five years. The Pacific Railway Act opened land and provided resources to complete the Transcontinental Railroad.

Ignoring that the territory was already occupied, however sparsely, our Euro-American ancestors faithfully followed what I've come to name in sermons as "the god of go take." From Father Abraham immigrating to Ur, to Moses leading his people to the Promised Land, was there ever any concern that these god-sanctioned *go to* places were already occupied?

Within two and a half years of signing the Pacific Railroad Act, Lincoln had "ethnically cleansed" the Dakota from Minnesota, the Cheyenne and Arapaho from Colorado, and the Navajo from New Mexico. Then, in 1864, he approved the creation of an Indian reserve for the Navajo people that, for all intents and purposes, served as a death camp. Over 10,000 Navajo men, women, and children were forcibly marched there. Hundreds died in route, and once there, those who attempted to escape were shot. Do we dare to wonder what the living conditions were like?

It should come as no surprise, then, that when he spoke about the German need for "living space," Hitler had America in mind. The Holocaust was the result: if Germany were to expand to the East, where millions of Jews lived, those Jews would have to vanish…because the Germans could not co-exist with them. No wonder that in *Mein Kampf* he praised America as the one state that had made progress in creating a racial concept of citizenship by excluding certain races from naturalization.

But such so-called *progress* doesn't necessarily go as planned. Case in point: when the railroad was scheduled to cross northern Nevada on the way to and from the California gold needed to underwrite the Civil War, the Western Shoshone neither sold nor ceded their tribal land for access. Instead, they signed a treaty of friendship with the U.S. Thus, legally they still own a large swath of Nevada... the very areas where the U.S. government has put its nuclear bomb test site and keeps trying to open a nuclear waste depository. Because the Shoshone never accepted the money the government offered for their land, they still claim that land and continue to act as a "stone in the shoe" to tweak the nation's conscience as it persists in seeing and using Nevada as a Wasteland.

But just as a grain of sand irritating an oyster becomes a pearl, the collective shadow can be more than *gifft*: the German word for "poison." For deep within our personal and collective Shadow lies a *gift* of great price, a pearl within the muck that's waiting to be faced and embraced. The worldview of peoples indigenous to this continent is a desperately needed corrective to the Western civilization mindset in which the Earth is a resource to be used and abused. Rather, they show us that Earth is a life-source to be cherished, celebrated, and protected.

The *gift* we've projected onto our indigenous peoples to hold onto for us, in spite of us, is the abiding wisdom to be found within our shared land, air, and water. Our numinous fascination with indigenous rites and rituals such as sweat lodges and vision quests invites us to cultivate our own authentic relationship to the land, rather than appropriating and coopting theirs. I like to think that we have begun to appreciate this continent where, according to Native American writer Vine Deloria (Standing Rock Sioux), God is Red.... and Mother Earth is the measure of all things, by setting aside public lands to be cherished and preserved as national parks, monuments, forests, deserts, watersheds, and wilderness areas to be protected against the relentless onslaught of profiteers.

Another opportunity to align ourselves with indigenous wisdom came about at Standing Rock Spirit Camp, set up to prevent the trespass of the Black Snake (oil and gas pipeline) across eco-sensitive indigenous lands. Capitalizing on our still colonializing dominant mindset and backed by the fossil-fuel entrenched federal government, the extractive industry went after these Water Protectors with attack dogs, water cannons, pepper spray, rubber bullets, and finally tanks to clear the way for the Dakota Access Pipeline to invade sacred tribal lands.

Before the camp was forcibly closed and bulldozed and resisters arrested,[41] veterans of the U.S. military flocked to the site to defend the water protectors and apologize for the past atrocities carried out by their cavalry forebears. Clergy from multiple faiths gathered in a circle around indigenous elders and ceremoniously burned a copy of *The Doctrine of Discovery*. Apologizing for harm done and promising to change the harmful behavior can heal the past in order to move into a hopeful future. Through making amends to the people we've wronged, the clergy, veterans, and hundreds of allies who spent time with the several tribes gathered at Standing Rock came away sensing they had received more than they had given: a life-changing experience of the sacredness of place.

Atonement becomes at-one-ment when we let ourselves feel interconnected with all life on this planet. One way to begin this shift in perspective may be to revise our Thanksgiving tradition of re-enacting the Pilgrims' giving thanks to their god that some of them survived their first harsh winter and had grown enough crops...thanks to the help of their native-born neighbors...to sustain them through the coming one.

In the national narrative we like to lift up once each year, the Pilgrims invited the Indians to their feast. Yet while I was serving my first church, a struggling congregation south of Boston and near Plymouth, I came across a 1996 book on *Native American History* by Judith Nies that claimed the local Wampanoags were celebrating their annual harvest festival and invited the Pilgrims to join them. This makes historical sense because the small band of simple-living religious dissidents, who were separating from rather than trying to purify the Church of England, would most probably have given thanks by fasting, not feasting.

After retelling that new perspective during every Thanksgiving worship service throughout my ministry career, I'd invite congregants to reconsider how they commemorated the nation-wide holiday: could we honor the native peoples' tradition of giving thanks *to* the fruits of the Earth, through the beans and squash, corn and cranberries? Might gratitude become the ladder by which we can climb up out of the abyss of our troubled conscience still hiding in our collective unconscious?

By celebrating Earth's harvest in an act of communion rather than as a re-enactment of colonialism, we just might start to redeem our national Shadow, and gain the needed gift of wisdom offered by the people we once sought to eradicate!

SIX. INDIGO
FAITH

"If we approach nature and the environment without openness to awe and wonder, if we no longer speak the language of fraternity and beauty in our relationship with the world, our attitude will be that of masters, consumers, ruthless exploiters, unable to set limits on their immediate needs."

—Pope Francis

WAS IT JUST A COINCIDENCE THAT, on the day I began reading the Pope's Encyclical on Climate Change, a mysterious little bird appeared in the aspen beyond the bay window of the Denver home? I've always loved lawn statues that show Saint Francis, his namesake, with a bird on his shoulder. How fitting that I would be able to watch an unfamiliar species build a nest in the battered old birdhouse that had never been used before. Where had it come from, this little remnant dinosaur that was probably far beyond its normal habitat range? Why was it here now? Would there be eggs? If any eggs even hatched, would there be the right foodstuff for chicks in our summer-scorched yard?

Wrapped in fascination at the unfolding eco-drama, I settled myself on the couch, computer propped up on my chest, and immersed myself in the online English version of the *Call of the Earth; the Cry of the Poor*.[42]

'Praise be to you, my Lord, through our Sister, Mother Earth, who produces various fruit with colored flowers and herbs.' In the words of this beautiful canticle, Saint Francis of Assisi reminds us that our common home is like a sister with whom we share our life and a beautiful mother who opens her arms to embrace us.

Squinting a bit, I could almost glimpse St. Francis standing on the other side of the bay window amid the columbine, under the aspen with its occupied birdhouse, preaching *to* nature, praising God.

Praise be You my Lord with all Your creatures, especially Sir Brother Sun, who is the day through whom You give us light. Praised be You, my Lord, through Sister Moon and the stars, In the heavens you have made them bright, precious and fair.

This form of prayer was perhaps inspired by an unusual encounter and interchange during one of the so-called *Holy*

Crusades. Francis, a lowly monk committed to peace, had become so disgusted with the carnage of endless wars against the "infidels" that he traveled behind enemy lines to meet with the Sultan. Over several days, Francis preached the case for peace, while simultaneously listening respectfully to the Sultan's Islamic perspective. As fate would have it, a bloodthirsty Christian bishop pushed for a victory that would let him brutally slaughter the opposition...but the Sultan won the battle. And this "infidel" refused to do unto the other side what they would have done to his troops. Had he been inspired by Francis' version of Christian witness? In return, was Francis inspired to adopt and adapt the structure of a Muslim prayer when he wrote, "Praise be to you, my Lord, through our Sister, Mother Earth, who produces various fruit with colored flowers and herbs?"

Now today's Francis has brought this forward in an Encyclical:

> This sister now cries out to us because of the harm we have inflicted on her by our irresponsible use and abuse of the goods with which God has endowed her. We have come to see ourselves as her lords and masters, entitled to plunder her at will. The violence present in our hearts, wounded by sin, is also reflected in the symptoms of sickness evident in the soil, in the water, in the air and in all forms of life. This

is why the earth herself, burdened and laid waste, is among the most abandoned and maltreated of our poor; she "groans in travail" (Rom 8:22). We have forgotten that we ourselves are dust of the earth (cf. Gen 2:7); our very bodies are made up of her elements, we breathe her air and we receive life and refreshment from her waters. [42]

Only in my imagination—so as not to bother the busy bird—the Pope joined St. Francis out in the garden while I read:

I do not want to write this Encyclical without turning to that attractive and compelling figure, whose name I took as my guide and inspiration when I was elected Bishop of Rome. Saint Francis is the patron saint of all who study and work in the area of ecology, and he is also much loved by non-Christians. He shows us just how inseparable the bond is between concern for nature, justice for the poor, commitment to society, and interior peace. [42]

I sat straight up on the couch, set aside the computer, and stared: Ralph Waldo Emerson had joined the Pope and the Saint in the garden.

Back after high school when I entered a hospital-based nurses training program, I didn't get to study Emerson's work as American literature like others do in college. Instead, I was in my first semester of seminary some thirty years later when I finally had a class for which I read his work deeply in its entirety, not just as excerpts out of context. That was because, before he was a renowned essayist and popular lecturer, Emerson was a Unitarian minister. His Transcendentalist wisdom came down to us as a major source of our living Unitarian Universalist tradition that today values the "direct experience of the transcending mystery and wonder, affirmed in all faiths, that calls us to a renewal of the spirit."

As Emerson described this experience in his essay *Nature*, "Standing on the bare ground, my head bathed by the blithe air and uplifted into infinite space, all mean egotism vanishes....the currents of the Universal Being circulate through me. I am part or particle of God."

The Pope quoted St. Francis as if to confirm Emerson's insights.

> If we feel intimately united with all that exists, then sobriety and care will well up spontaneously. The poverty and austerity of Saint Francis were no mere veneer of asceticism, but something much more radical: a refusal to turn reality into an object simply to be used and controlled. [42]

Emerson nodded his appreciation and affirmation, then they began a dialogue across centuries and languages and locations.

Before I could blink, St. Francis insisted on including the Muslim perspective on the growing ecological catastrophe that is coming to pass today as humans change the climate. Smiling his agreement, Pope Francis invited a representative from the 2010 International Conference on Muslim Action on Climate to share their statement:

> Preservation of the earth's ecosystem is the preservation of life. Global warming is mainly caused by the continuous increase in human consumption driven by a global paradigm that is anthropocentric and has economic growth as its primary objective. The holistic Islamic concept of the blessing of the universe necessitates that we share the world fairly with all mankind through sustainable development. We believe that global good environmental governance can be achieved with the Islamic principles of balance, leadership, stewardship, and collaboration.[43]

Emerson, in his abiding fascination with eastern religions, naturally insisted upon bringing in the perspectives of Hinduism and Buddhism:

In the Hindu spirit of 'the whole world is one family,' we cannot continue to destroy nature without also destroying ourselves. We can and should take the lead in Earth-friendly living, personal frugality, lower power consumption, alternative energy, sustainable food production and vegetarianism, as well as in evolving technologies that positively address our shared plight. We must do all that is humanly possible to protect the Earth and her resources for the present as well as future generations.[44]

We are violating the first Buddhist precept—do not harm living beings—on the largest possible scale. Today we live in a time of great crisis, confronted by the gravest challenge that humanity has ever faced: the ecological consequences of our own collective karma. The scientific consensus is overwhelming: human activity is triggering environmental breakdown on a planetary scale. We have a brief window of opportunity to take action, to preserve humanity from imminent disaster and to assist the survival of the many diverse and beautiful forms of life on Earth. We must listen to them and the silence of future generations. We must be their voice and act on their behalf.[45]

Then the Pope, before adding his own climate encyclical, added the wisdom of the other Abrahamic faith, Judaism:

> Our tradition teaches that Adam and Eve were asked 'to till and to tend' the Garden of Eden, and we believe humans remain a partner in Creation. We fulfill this mandate by practicing 'Tikkun Olam,' literally repairing the world. Because climate change threatens to irreparably alter the Earth, the organized Jewish community is united in its deep concern that the quality of life and the earth we inhabit are in danger. We affirm our responsibility to address this planetary crisis in our personal and communal lives in order to protect the Earth for future generations.[46]

Wait, wait, I wanted to shout, before we get back to the Pope's brilliantly inclusive statement, may we please lift up Indigenous wisdom as a foundational part of our human response-ability towards our shared natural world? I felt compelled to do this because the sources of my faith tradition include "Spiritual teachings of earth-centered traditions which celebrate the sacred circle of life and instruct us to live in harmony with the rhythms of nature."[47]

This source was adopted at the annual assembly I missed the summer I was doing my chaplaincy training in Billings, Montana, at a hospital that served nearby Indian populations.

Among other experiences, I was assigned a comatose Crow Indian. Morbidly obese, he had heart disease and severe diabetes so his prognosis was poor, and it was my duty to get the family to pull the plug and let him go. Instead, I met with his wife as she sat guarding the door to his room, and listened to whatever she wanted us to know. The family wasn't about to make any decision until after the upcoming Sun Dance. A designated participant was to pray for an answer in a vision at the event.

I traveled down to the Crow Reservation on the day of the Sun Dance. Having spent many hours there over the summer for various reasons, I knew my way around. Plus, a fellow chaplain, a Crow woman, had even held a sweat lodge ceremony for the women student chaplains. I relished the drive across the wide-open plains where the horizon was as circular as a teepee. The boundless sky was an acquired taste to be sure, and it took a long while to push past the weak knees and queasy stomach that came from having such an unobstructed view, with no point of reference to anchor my body, mind, and spirit. I've heard that early pioneers compared crossing the plains in covered wagons to crossing the ocean in ships. Was that why they were called "prairie schooners?"

Meanwhile, I'd learned to listen to the earth in a whole new way, so when the patient's wife revealed the dancer's vision that her husband was to live, I could, and did, support her truth against the pressure of the rest of the hospital staff and administration, worried over wasting resources.

After centuries of trying to wipe out Indian customs and culture, thus propelling our native peoples into the self-destructive lifestyles that result in alcoholism, diabetes, hopelessness and depression, the legal reinstating of the Sun Dance and other cherished rites was a critical step in honoring native wisdom. No wonder that today we find our American Indian tribes at the forefront of the climate struggle. This was affirmed when I found the following in a booklet on *Faith Statements on Climate Change* composed by two members of the Citizens Climate Lobby.

> Mother Earth is no longer in a period of climate change, but in climate crisis. We are deeply alarmed by the accelerating climate devastation brought about by unsustainable development. We reaffirm the unbreakable and sacred connection between land, air, water, oceans, forests, sea ice, plants, animals and our human communities as the material and spiritual basis for our existence. We therefore

insist on an immediate end to the destruction and desecration of the elements of life.

Through our knowledge, spirituality, sciences, practices, experiences and relationships with our traditional lands, territories, waters, air, forests, oceans, sea ice, other natural resources and all life, Indigenous Peoples have a vital role in defending and healing Mother Earth. We offer to share with humanity our Traditional Knowledge, innovations, and practices relevant to climate change, provided our fundamental rights as intergenerational guardians of this knowledge are fully recognized and respected. We reiterate the urgent need for collective action. [43]

Finally, as if gathering up everyone's input and concerns, the Pope summarized his own message on climate change:

The earth, our home, is beginning to look more and more like an immense pile of filth. The human environment and the natural environment deteriorate together; we cannot adequately combat environmental degradation unless we attend to causes related to human and social degradation.

The warming caused by huge consumption on the part of some rich countries has repercussions on the poorest areas of the world. Such effects will continue to worsen if we continue with current models of production and consumption.

We are faced not with two separate crises, one environmental and the other social, but rather with one complex crisis which is both social and environmental. Strategies for a solution demand an integrated approach to combating poverty, restoring dignity to the excluded, and at the same time protecting nature.

Today, in view of the common good, there is urgent need for politics and economics to enter into a frank dialogue in the service of life, especially human life.[42]

Months after I'd read and studied the Pope's *Encyclical*, and with the 2015 UN Climate Summit coming up in Paris as a make-or-break opportunity to take action worldwide before times runs out, Pope Francis traveled to the United States to address a joint session of Congress.

Internationally, our intractable federal government has been the foremost obstacle to global climate action for decades, in large part thanks to the campaign funding of corrupt politicians by powerful fossil fuel corporations. Although I am a vocal advocate for the separation of church and state that the founders incorporated into our Constitution, I had to appreciate, and applaud, the Pope's determined commitment to speak of climate change's causes and consequences, along with the wider faith community's call for a moral response through ethical actions.

> In Laudato Si', I call for a courageous and responsible effort to 'redirect our steps,' and to avert the most serious effects of the environmental deterioration caused by human activity. I am convinced that we can make a difference and I have no doubt that the United States—and this Congress—have an important role to play. Now is the time for courageous actions and strategies, aimed at implementing a culture of care and an integrated approach to combating poverty, restoring dignity to the excluded, and at the same time protecting nature.[48]

The Pope was repeatedly applauded by the Democrats, many of whom had recently pulled an all-nighter on the Senate floor, talking about what each of their states was doing to deal with climate change, but the die-hard Republicans, even those claiming to be practicing Catholics, blew off the Pope's remarks and insisted he should stick to religion and leave science to the scientists. Apparently, they forgot, or chose to ignore, that the Pope studied chemistry in undergraduate school!

People of faith who are determined to walk their talk are not so easily dismissed or dissed: beliefs frame worldviews and shape behavior.

The summer of the Pope's Encyclical, as we watched the parent birds struggle, and fail, to keep the chicks in the birdhouse alive, Milt and I prepared to put our Unitarian Universalist climate statement into action:

> We are part of this world and its destiny is our own. Life on this planet will be gravely affected unless we embrace new practices, ethics, and values to guide our lives on a warming planet. We will not acquiesce to the ongoing degradation and destruction of life that human actions are leaving to our children and grandchildren. We envision a world in which all people are assured a secure and meaningful life that is ecologically responsible and sustainable, in which every form of life has intrinsic value.
>
> Our world is calling us to gather in community and respond from our moral and spiritual wealth; together we can transform our individual and congregational lives into acts of moral witness, discarding our harmful habits for new behaviors and practices that will sustain life Earth, ever vigilant against injustice.[49]

Determined to embrace the principle of our faith that commits us to live with radical respect for the Interdependent Web of all existence of which we are but a part, we committed to stop being part of the problem and pushed for a clean energy system in the building renovation being planned for Milt's Universalist church. A lot of footwork was needed first.

When I was invited to give a sermon on climate change, I challenged the claim that promoting environmental values somehow threatened jobs by pointing out that our English words ecology and economy both come from the root word ecos, Greek for "home." Then I went on to suggest that any economic activity that threatens the ecosystems on our home planet is insane, immoral and ought to be deemed illegal. For just as the cotton-based economy of the civil war era did not justify slavery, the fossil fuel-based economy of today does not justify enslaving the living Earth. That's why, as ordinary people, we are taking action: we are banning fracking, we are blocking pipelines, we are lobbying Congress for a carbon tax, we are divesting our personal and professional portfolios from fossil fuels, and we are demanding that our churches and colleges do likewise.

This worship service was followed by a workshop that offered specific tools for living eco-responsibly beyond just changing light bulbs. We also needed to change our life ways by weaning ourselves off our fossil fuel addiction and insisting on renewable, non-polluting ways to empower our lives, such as installing rooftop solar to power homes and electric cars, sinking geothermal loops for our heating and cooling, growing and/or eating local, organic foods, and driving fuel-efficient cars and/or biking. In short, we set forth actions that took us beyond our planet-destroying consumerist habits.

During the discussion period that followed, one of the 35 attendees remarked that she wouldn't pledge to the capital campaign being put in place to finance the church renovation unless it included a renewable energy system. When others in the workshop concurred, the stage was set for the church's Green Sanctuary Team to push for the new facility to be carbon neutral, and find ways to make it happen.

This intrepid little group had its roots in the Unitarian Universalist association-wide Green Sanctuary Movement, an effort to make individual congregations more mindful of their ecological interconnectedness and responsibility via worship services and religious education programs as well as by energy audits and retrofits of the church buildings. The challenge of a clean, renewable energy system was a logical next step.

Yet logic couldn't compete with the reality of raising funds for the new build when the religious ideal of protecting the future habitability of the planet collided with the practical need for more classroom space for the children of today. One of the first things on the congregational wish list to get cut when the capital campaign came up short was the geothermal part of the clean energy system. Solar could be installed through a third-party lease agreement, but money for geothermal was *not* going to come out of the building fund.

Now what? Giving up was not an option for the Green Sanctuary Team. Even though the cost of a clean energy system was less than ten percent of the cost of the remodeling project, the board of trustees sided with the building committee charged with working out the details and funding the entire 4-million-dollar project. As it was, the church would have to take out a commercial bank loan to cover the $400,000 not raised during the capital campaign.

The Green Sanctuary Team offered to raise the funds for the renewable energy system. The board indicated that any proposed funding approach for the new energy system would have to be "revenue neutral," meaning that there could be no change in the church operating budget.

After months of wrangling over how to fund the total clean energy system, the Team finally came up with the idea of using low-interest member loans to fund the solar electric and geothermal heating and cooling systems as a total renewable energy system with zero greenhouse gas emissions. These loans would be serviced out of the church's current operating budget—instead of paying the utility company monthly for electricity and natural gas, the church would pay-down the members' loans every month. With heating and cooling costs zeroed out by producing its own energy, the church budget would not have budged.

At a congregational meeting, church members gave the go-ahead for the total renewable energy system. Most of the faithful church donors were already tapped out, so the Green Sanctuary Team created a "partnership" and recruited lenders and new donors. In the end, the necessary capital was raised, thanks to fifteen church members who loaned $240,000 at 1.5% interest for 15 years, and forty others who donated $200,000 for the clean energy system.

The church renovation proceeded, with the usual set-backs and delays and financial over-runs expected of any major build. The finished building was not only beautiful, with its stunning sanctuary ceiling of reclaimed bark- beetle-kill wood, it was also sustainable, and wouldn't harm current and future generations.

Inadvertently, the church became a model for what was possible! Workshop opportunities presented themselves, each to be concluded with a tour of the area where the geo-thermal pipes and pumps were located. Interested groups escorted down into the church basement included the Colorado Renewable Energy Society, as well as congregants and visitors from other churches. The church became the obvious place to hold a local event for the Rise for Climate global action day being held in conjunction with the Global Climate Action Summit, a gathering of leaders from cities, states, businesses and civil society from around the world that invited every mayor, governor, and local leader in the world—whether they were at the summit or not—to make a bold climate commitment to help the world reach the goals of the Paris Climate Agreement.[50]

Our workshop emphasized the urgency of the need to achieve a fast, fair and just transition to 100% renewable energy and an immediate end to new fossil fuel projects. It was followed by a tour of the clean energy system. Milt had painstakingly painted the basement area and cleared away the stored stuff no one knew what else to do with, then decorated the passageway walls with posters depicting the science behind global warming and the mechanics of the geothermal system.

To this *what* and *how* I added the *why*...by creating a mural on one white washed wall. It featured the famous photo of Earth from space, surrounded by the climate statements of all the major faith traditions.

SEVEN. VIOLET
FUTURE

"Youth shouldn't have to throw away our childhoods
and devote our lives to convincing our leaders to let
us breathe clean air, drink potable water, and live on
a habitable planet."

—Denver Kids Climate Summit

WHAT DO WE TELL THE CHILDREN? They should be
playing with toy dinosaurs and searching for fossils here in
Colorado instead of worrying about following them into
extinction. How tragic, then, that an eleven-year-old boy
from Boulder, on his way home from an environmental conference in 2014, suddenly asked his mother, "Why should I
go to school and learn a bunch of stuff if there is not going to
be a world worth living in?" Then he stopped talking.

Deciding to be "silent to be heard," he set off a worldwide silent strike on social media in a desperate attempt to
bring attention to the changing climate's threat to his future.
Thousands of children and adults signed on to join him for
a day or hour of silence in mid-December, when the United
Nations met in Lima, Peru, to prepare for the major climate

talks the following year in Paris, a meeting seen to be the last best chance to turn things around before the climate-action window closes for good.

We all wore earth-blue wristbands to symbolize our solidarity in insisting that world leaders take concrete action on this intergenerational crisis. This youngster, along with his older brother, had early on become activists in a group started by their big sister: Earth Guardians is an "international youth organization committed to protecting the Earth, water, air, and atmosphere so that their generation and those following can live on a healthy, just, and sustainable planet."

What do we tell the children who have every right to expect that the adults responsible for them will ensure they have a sustainable planet to grow up on and a stable climate now and into their future?

The climate-denying industry supplies an answer. They've created and distributed junk science-filled textbooks to every teacher and school across the country, claiming the climate has always changed and that humans can't do anything about it.[51] Children are not to worry about the dire warnings from radical environmentalists who want to destroy the economy and kill off everyone's right (and obligation) to consume more of the stuff these polluters produce for our pleasure and their profit.

Don't we need to teach our children that real science doesn't lie and that there's an observable correlation between cause and effect? Dumping greenhouse gases into the atmosphere thickens the protective blanket enveloping the earth too quickly for the planet's species, including people, to adapt or mitigate or reverse. Shouldn't adults set out the evidence and then model reason, exercise critical thinking, and use common sense to solve the problem? Afterall, that's why President Thomas Jefferson was so adamant that citizens in a democracy must be educated, and founded the University of Virginia for the public instead of the privileged.

Children across the world who *have* made the scientific connection between climate cause and effect are filing lawsuits against national and local governments for violating their right to a livable planet.

Meanwhile, in the 2015 Paris Climate Agreement, our own government finally committed to reducing our emissions, with a national clean power plan as the major mechanism. But now the current administration has withdrawn us from that agreement and not only reversed all efforts to move to renewable energy but actually ratcheted up the exploration and exploitation of dirty energy sources. Burning these fossil fuel reserves is already increasing our carbon emissions by an unconscionable, irreversible amount. With no globally binding commitment to reduce global emissions, extreme weather events have been locked in for today's children and *their* children and grandchildren.

This is the crime against humanity that today's children are up against. They've given up expecting the legislative and executive branches of our government to do their constitutional duty, and turned to the judiciary branch. In what feels like a final act of faith in democracy, twenty-one young people are suing the administration under the public trust doctrine that compels the government, as trustee, to protect the country's natural inheritance for current and future citizens.[52]

The children have standing to bring the case because they will be more adversely affected by climate change than any adult now living. Their legal claim is also based on the constitutional doctrines of Due Process and Equal Protection. Sixty-three professors of law have signed on to the amicus brief filed in support, as well as have numerous citizens groups, including the one I created of, for, and by eco-elders.

That governments have environmental obligations to generations to come is now a feature of international treaties, national constitutions, and case law. Our government has known since the 1960s that burning fossil fuels was causing the earth to warm, would disrupt the climate, and trigger extreme weather events, but chose to do nothing about it. Instead they opted for stalling not solving, for denying and doing nothing. Our leaders essentially kicked the can down the road for future generations to deal with. These are the children of that future.

Who are the children currently challenging our government and demanding action to keep the atmosphere at no more than 350 parts per million of CO_2 and to no more than one degree of warming, instead of the 450 ppm and 2-plus degrees predicted if we continue on the do nothing-keep-polluting business-as-usual trajectory?

The lawsuit itself bears the name of a now 22-year-old who's been engaged in climate activism since she was 10, because "I believe that climate change is the most pressing issue my generation will ever face, indeed that the world has ever faced. This is an environmental issue and it is also a human rights issue." She was recently recognized as an addition to the collection of American's Who Tell the Truth.[53]

The youngest plaintiff, at 11 years old, has been honored by my own faith tradition's Ministry for Earth as a Guardian of the Future. Living on a barrier reef off the coast of Florida that has been impacted by toxic red algal blooms, increasing storms, and rising sea levels, he rebuilds sand dunes as he grieves the loss of nesting sea turtles.

Another plaintiff attends the same college in Pennsylvania as my second granddaughter. Growing up with the stories of climate change from her legendary grandfather, Dr. James Hansen, she became passionate about the climate at a young age. Frequently missing school because of extreme weather, she grew deeply concerned that climate change will not only harm her, but will also harm the entire fabric of human civilization and all living things on Earth that she cherishes and relies on for her life, liberties, and prosperity. "Our government isn't taking action and we have a very small window to turn this around. I feel a sense of responsibility to take action," she says on Our Children's Trust's website.

I unexpectedly got to meet a youth plaintiff here in Denver when he joined his mother, grandmother, and me for a lunch meeting on faith communities and climate change. Still in high school, he is drawn to the intersection between church and environment, and is especially inspired by Pope Francis' climate encyclical. "Our moral call to protect the common good (and not the profits of a selfish few) is clear in the Pope's call to action." He has taken that command to heart, and is acting to "protect our planet and our civilization."[53]

The suit was first filed in 2015, with a complaint that the government's actions and inactions around climate change have violated the youngest generation's constitutional rights to life, liberty, and property, as well as failed to protect essential public trust sources.

The fossil fuel industry initially intervened in the case as defendants, joining the U.S. government in trying to have the case dismissed, but then withdrew when it became clear they would be required to hand over all documents showing how, when, and what they knew about their product's effect on the climate. (Since then, ExxonMobil has been required to turn over files for a similar court case in Massachusetts.)[54]

Dismissal requests from the government have been heard and then rejected by successive judges for over three years, thus delaying the trial. In the summer of 2018, the U.S. Supreme Court unanimously ruled in the children's favor by denying the current administration's application for stay in a last ditch effort to stop the trial scheduled for October 29, 2018 because "the right to a stable climate system capable of sustaining human life is fundamental to a free and ordered society." By October, the administration had filed for yet another stay, had it denied, appealed the ruling in November, had it accepted, and the trial was delayed yet again.

While we wait, and wait, and wait, it is helpful to compare this case with the Supreme Court's Dred Scott decision in 1857 that ruled against the defendant because slaves and their descendants were "beings of an inferior order" who had "no rights which the white man was bound to respect." As such they had no standing to sue for their freedom in federal court. Thirteen years later the 13th Amendment abolished slavery because, by then, the Dred Scott ruling was out of touch with the growing public awakening as well an evolving economic reality.

Are the rights of future Americans to a livable environment likewise setting into motion a precedent that can't and won't be ignored? Or will the politically influential fossil fuel industry be allowed to hold onto the past long enough to extract every last ounce of profit from their product before they push these children, as the face of humanity, off the cliff of extinction? Perhaps the moral arc of history will bend towards the justice of no longer allowing today's citizens to diminish the lives of tomorrow's children by neglecting to shift to safer, saner energy sources.

While we're waiting for the trial to finally happen, the latest Congressionally mandated national climate assessment was released on the slowest news day of the calendar year: Black Friday. Those of us boycotting this annual spending frenzy used that time to study its 1600 pages prepared by 300 scientists affiliated with 13 government agencies.

It came on the heels of the latest, most devastating yet, report from the U.N.'s Intergovernmental Panel on Climate Change, and while the worst wildfire in California's history was still sending us smoke signals. There's a mind-boggling disconnect between its warning and the carbon emissions being unleashed as millions of shoppers spend a trillion dollars fighting one another to secure the latest media-hyped must-have present.

Yet tracing the carbon footprint of anything containing plastic, for instance, reveals a path of community decimation and environmental degradation, along with the increase of CO_2 in the atmosphere. Oil is extracted via fracking that pollutes neighboring air and water. It is piped through fragile ecosystems to be processed into plastic via toxic chemicals that poison workers before the waste is pumped into public waters. Then the plastic is shipped abroad to be manufactured into cheap stuff by underpaid employees in environmentally deregulated countries. Sent back to our country by ships that run on highly polluting bunker fuel, and trucked to superstores or stacked in warehouses serviced by workers not allowed bathroom breaks, these products are wastefully packaged in cardboard so they can be delivered to the doorsteps of consumers. The total carbon emissions generated by this process will severely degrade the future of the children who are receiving the Christmas gifts, thus making the case for the children's lawsuit.

In a heart-wrenching speech at COP24 in Poland, a 15-year-old Swedish girl named Greta claimed that because national leaders have failed to act for 25 years, she was not going to ask them for anything. Instead, she asked the media to start treating the climate crisis as a crisis. She asked the people around the world to realize that the political leaders have failed and now "we are facing an existential threat and there is no time to continue down this road of madness. And since our leaders are behaving like children, we will have to take the responsibility they should have taken long ago."[55]

It's hard not to see this young girl as the modern incarnation of St. Lucia, the legendary Swedish light-bringer during the season of darkness. For months this child sat on the

Swedish Parliament steps instead of going to school. "Why should I prepare for a future that might not happen? Why should I study facts when clearly facts don't matter?"

As she recently explained in a TED^X Talk in Stockholm, she was 9 years old when she first learned about climate change and became so depressed that she stopped eating, and then speaking. At age 11, she was diagnosed with Asperger's Syndrome, OCD, and "selective mutism," "which means that I speak only when I have something to say."

She has a lot to say about climate change.

> Our emissions are still increasing at the same time the science has clearly told us that we need to act now to keep the planet within 1.5 degrees of warming. Whoever you are, wherever you are, we need you now to stand outside your parliament or local government office to let them know that we demand climate action.
>
> Adults keep saying: 'We owe it to the young people to give them hope.' But I don't want your hope. I want you to panic. I want you to feel the fear I feel every day. And then I want you to act. I want you to act as you would in a crisis. I want you to act as if our house is on fire. Because it is.[56]

Greta Thunberg's continued weekly school strikes for the climate in her native Sweden would inspire children across Europe, and then the world, to do likewise. She would come to the United States by carbon-footprint-free sailboat, address Congressional committees and a United Nations special meeting on climate action, make multiple television appearances, and lead a massive school strike. Then she would travel across the country leading Friday school strikes and climate rallies before sailing to Spain to participate in the 2019 COP meeting.

When the adult in the room is a child, perhaps it is time to revisit the child within each of us...maybe that one in pajamas who's standing in front of the Christmas tree surrounded by gift-wrapped surprises, filled with anticipation of fulfilled wishes. Back then these presents were as concrete as dolls and dinosaurs, ice skates and sleds. As adults they're abstractions wished for us on Christmas cards: hope, health, happiness, love, peace, joy. All reflect the best that the universe holds for our species, made manifest in the spirit of children. Children of today wish and work for a sustainable world and stable climate.

The winter solstice season has always been a time to reclaim a child-like trust in spring's promise that's already budding along barren tree branches, inviting us to hope again. Resolving to try, I set up our Charlie Brown tree again, this time fastening photos of my five grandchildren and Milt's three step great-grandchildren onto its scraggly branches. Beneath this solstice altar, I placed a plush green gas station mascot dinosaur to symbolize the last mass extinction, and the fossil fuel cause of the coming one. Beside it I set a red-sweater-wearing polar bear to signify the 200 species going extinct daily at human hands.

My daily meditations by candlelight in the sacred morning darkness focused on the question of what do we say to *all* these children, we grown-up human children who are "old enough to know better?"

Another election cycle *has* swept out nearly forty old-school seats and opened up space for something new through a diversity in gender, ethnicity, and age that more faithfully represent we the people. One hundred women have taken their rightful place at the leadership table, including two Native Americans. The youngest new congresswoman, influenced by the stand-off at Standing Rock where she witnessed "others putting their whole lives and everything they had on the line to protect their community," has provided a platform for the Sunrise Movement.[57] This rapidly growing grassroots group of young adults has organized to advocate for political action on climate change, and is championing a Green New Deal that would switch the energy infrastructure of the U.S. from fossil fuels to renewables, while creating millions of jobs in the process. Their plan demands the move to 100% renewable energy within the exact time-frame that scientists say we have left to act.

Nearly immediately, opponents began screaming that such a transition is impossible so fast and would cost trillions of dollars. Even would-be supporters warn that such a plan is too radical. But slow and steady is no longer an option! We've denied and delayed long enough! Had we paid attention and acted responsibly all along the way, we wouldn't have to take such drastic corrective measures today. Now there's no choice.

Embracing this glimmer of hope, I unpacked the decorations for the rest of the house, removed my delicate Nurnberg Angel from her little box, and let her trigger my memory of visiting the famous *Nuremberg Christkindlesmarkt* with my own young daughters back in 1978.

Suspended in flight and presiding over the outdoor array of booths brimming with seasonal goodies for sale, the signature Nuremberg Angel is a beloved symbol of childhood innocence reflecting two popular legends. In one, a grief-stricken mother fashioned a little angel and shaped wax into a facsimile of the face of her dead daughter. Then she hung the angel on her Christmas tree. When neighbors saw it, they too wanted such an angel to grace their trees. Thus, private grief became a public treasure in an age when people were desperate for beauty amidst the destruction of endless war. She began to mass-produce her delicately-featured angels with their golden foil wings. In another legend, it was a toymaker who, as his daughter lay dying of a fever, heard the flutter of angel's wings, and was inspired to create an angel in her memory.

Since then, the Nurnberg Angel has come to symbolize the innocence and magic of childhood. Yet this sentiment isn't always or easily translated into the life of real children. For instance, when my own small daughter was being elbowed away from the *Christkindlesmarkt* stalls by heavily coated matrons determined to finish their shopping, she was forced to take the matter into her own hands and came up with a way to move through the crowd. She pretended she was going to throw up, and the sea of people parted. I was horrified as I followed behind her, but grateful to make it to the stall where we selected a small Nurnberg Angel. Clothed in maroon velvet and white chiffon, with gold foil wings, her wax face has smiled sweetly from the top of our Christmas trees down through the years. Now I placed her on our little tree along with those faces of the future calling on us to stop trashing the planet and crashing the climate.

What do I say to the children?

"I'm sorry" rings as hollow as a drunk slobbering regrets between binges who succumbs to the addiction again when gas prices drop and everyone can continue to compromise the planet with our care-less-ness.

Instead, I made a New Year's wish for all the children of today that, rather than trying a crime that's already been perpetrated against humanity as happened after the Holocaust with the Nuremberg Tribunals of 1945-46—we grown-ups will help ensure that a trial on today's existential threat to humanity comes to pass *before* this crime fully happens, thus hopefully stopping this insanity before we pass the point of no return.

AFTERWORD

ANYONE *NOT HAVING NIGHTMARES* isn't paying attention. Climate experts update their predictions nearly weekly. Tipping points are being reached earlier than expected, and the point of no return is being breached with impunity and ignorance, arrogance and apathy.

Gloom and doom abound.

Our country has gone rogue as the only nation not even pretending to comply with the climate goals set by the international community. In fact, our CO_2 emissions have increased since the climate agreement was made, and are now set to offset the gains made by all the other countries combined. It's as if our nation's climate deniers have become climate destroyers, and are deliberately doubling down on putting power and profit above people and planet. Perhaps I should send them some plastic dinosaurs.

Exhausted from riding the roller coaster between hope and despair, I take a time out to sit still, pull my silly green brachiosaurus onto my lap, and whisper: "We humans have something you dinosaurs didn't: creative imagination! People *can* dream a different future, and then co-create the path into it. But only if we want to and do so."

Either we will get serious about the urgency of the situation and change our light bulbs, our life ways, and our leadership, or we might as well just give up, give in, and go the way of the dinosaurs.

Day and nightmares morph into liminal space, that time between what is and what is to be, with those of us lined up along the cliff-face of extinction still holding on, holding back, holding fast, holding open the place for transformation to happen.

The Sun still has our backs. The rainbow has doubled now, though the second one is almost too faint to be seen without faith that it's actually there. Squinting, I note that its colors are the reverse of the first, as if everything we've accepted as normative has been turned upside down. Which is exactly what has to happen! The red band of warning must be at the bottom, with the violet strand for the future arching over the whole, staking its claim as our top priority! Red warnings are stronger than ever; orange activism has been ignited; yellow sunlight and other renewables have become cheaper than fossil fuels; the political will to act is growing thanks to green-minded voters; indigenous peoples are still standing steadfast on behalf of the earth; more and more people of faith are aligning beliefs with behavior; and our precious, precocious, unstoppable children are striking for a future that is possible.

Maybe, just maybe, our climate moment will gather enough momentum to become the movement that can finally resolve this crisis and keep humanity from extinguishing itself.

ENDNOTES

ONE. RED: WARNING

"History of Climate Change Science," Wikipedia, accessed August 31, 2019,https://en.wikipedia.org/wiki/History_of_climate_change_science.

2 "Exxonmobil Climate Change Controversy," Wikipedia, accessed July 31, 2019, https://en.wikipedia.org/wiki/ExxonMobil_climate_change_controversy.

3 "The Carbon Dioxide Problem," *The New York Times*, accessed July 29, 2019, https://static01.nyt.com/packages/pdf/science/woodwellreport.pdf.

4 "Climate Change Denial," Wikipedia, accessed July 31, 2019, https://en.wikipedia.org/wiki/Climate_change_denial.

5 "A Climate Hero the Early Years," accessed July 20, 2019, https://grist.org/article/a-climate-hero-the-early-years.

6 "The Carbon Dioxide Problem," *The New York Times*, accessed July 31, 2019, https://static01.nyt.com/packages/pdf/science/woodwellreport.pdf

7 "United Nations Framework Convention On Climate Change," accessed July 31, 2019, https://en.wikipedia.org/wiki/United_Nations_Framework_Convention_on_Climate_Change

8 "Rio+20: Earth summit dawns with stormier clouds than in 1992," *The Guardian*, accessed July 30. 2019, https://www.theguardian.com/environment/2012/jun/19/rio-20-earth-summit-1992-2012.

9 "How Big Oil Lost Control of Its Climate Misinformation Machine," *Inside Climate News*, https://insideclimatenews.org/news/22122017/big-oil-heartland-climate-science-misinformation-campaign-koch-api-trump-infographic.

10 "The Halliburton Loophole," accessed July 30, 2019, https://earthworks.org/issues/inadequate_regulation_of_hydraulic_fracturing/

11 "Who Killed the Electric Car," accessed July 30, 2019, https://en.wikipedia.org/wiki/Who_Killed_the_Electric_Car%3F.

12 "National Geographic Bulletin from a Warmer World," accessed July 25, 2019, https://www.google.com/from+a+Warmer+World+National+Geographic+2004.

13 "Dark Money: Jane Mayer on How the Koch Bros. & Billionaire Allies Funded the Rise of the Far Right," *Democracy Now!*, accessed August 27, 2019, https://www.democracynow.org/2016/1/20/dark_money_jane_mayer_on_how

14 Global Warming—when Politics And Science Collide," accessed July 25, 2019, *Answers Magazine*, https://answersingenesis.org/environmental-science/climate-change/global-warming.

TWO. ORANGE: ACTIVISM

15 "ALEC Exposed," The Center for Media and Democracy, accessed July 29, 2019, https://www.alecexposed.org/wiki/ALEC_Exposed.

16 "Is the World Getting Hotter?", *The New York Review of Books,* accessed July 30, 2019, https://www.nybooks.com/articles/1988/12/08/is-the-world-getting-hotter/.

17 "Do the Math Tour," The Donella Meadows Project, accessed August 1,2019, http://donellameadows.org/do-the-math-tour/.

18 "350.org," accessed July 30, 2019, https://350.org.

19 "Bill Mckibben's Speech At Power Shift 2011," 350.org, accessed August 1, 2019, https://350.org/bill-mckibbens-speech-power-shift-2011.

20 "The Keystone Pipeline Revolt: Why Mass Arrests Are Justified," *Rolling Stone,* accessed July 31, 2019, https://www.rollingstone.com/politics/politics-news/the-keystone-pipeline-revolt.

21 "Top NASA Scientist Arrested (again) In White House Protest," accessed July 31, 2019, https://www.foxnews.com/science/top-nasa-scientist-arrested-again-in-white-house.

22 "Quote By Kate Angell: 'A Good Friend Will Come,'" accessed July 31, 2019, https://www.goodreads.com/quotes/536624-a-good-friend-will-come-and-bail-you-out.

23 "The Great March for Climate Action, Mission," accessed July 31, 2019, h http://climatemarch.org/the-great-march-for-climate-action/mission-statement/.

24 "2014 People's Climate March," Wikipedia, accessed July 31, 2019, https://en.wikipedia.org/wiki/2014_People%27s_Climate_March.

25 "Dakota Access Pipeline Protests," Wikipedia, accessed July 31, 2019, https://en.wikipedia.org/wiki/Dakota_Access_Pipeline_protests.

THREE. YELLOW: SUNLIGHT

26 "An Ill Wind Blows in Moapa," Earth Justice, accessed July 28, 2019, https://earthjustice.org/blog/2011-july/an-ill-wind-blows-in-moapa.

27 "Salutation to the Dawn," Belief Net, accessed July 31, 2019, https://www.beliefnet.com/prayers/hinduism/morning/salutation-to-the-dawn.aspx.

28 "Happening, A Clean Energy Revolution," The Redford Center, accessed August 15, 2019, https://happeningthemovie.com/about-the-film/.

FOUR. GREEN: BRIDGING

29 "Finally, Neil deGrasse Tyson and 'Cosmos' Take on Climate Change," Grist.org, accessed August 17, 2019, https://grist.org/climate-energy/finally-neil-degrasse-tyson-and-cosmos-take-on-climate-change/

30 "Fossil Fuel Subsidies Overview," OilChange International, accessed August 18, 2019, http://priceofoil.org/fossil-fuel-subsidies/.
31 "Gasland the Movie," accessed August 19, 2019, http://one.gaslandthemovie.com/home

32 "Theo Colborn," Alchetron, The Free Social Encyclopedia, accessed July 31, 2019, https://alchetron.com/Theo-Colborn.

33 "Anti-fracking Activist Asks Court To Lift Ban," RT, Question More, accessed July 31, 2019, https://www.rt.com/usa/fracking-injunction-pa-scroggins-893/.

34 "Neighborhoods Worry About Living Amid Oil And Gas," NPR, accessed July 31, 2019, https://www.npr.org/2017/08/26/545583191/neighborhoods-worry-about-living-amid-oil-and-gas-development.

35 "Parents Didn't Want Fracking Near Their School. So the Oil Company Chose a Poorer School, Instead," *Mother Jones*, accessed July 31, 2019, https://www.motherjones.com/environment/2018/04/an-oil-company-faced-pushback-about-fracking-near-a-charter-so-it-moved-next-to-a-low-income-public-school/

36 "Fracking Increases Risk Of Asthma, Birth Defects," *Rolling Stone,* accessed July 31, 2019, https://www.rollingstone.com/politics/politics-news/the-harms-of-fracking-new-report-details-increased-risks-of-asthma-birth-defects-and-cancer-126996/

37 "The Permanent Peoples' Tribunal on Human Rights, Fracking, and Climate Change," accessed July 31, 2019, https://www.tribunalonfracking.org.

FIVE. BLUE: SHADOW

38 "Carl Jung, the Shadow, and the Dangers of Psychological Projection," Academy of Ideas, accessed August 31, 2019, https://academyofideas.com/2018/02/carl-jung-shadow-dangers-of-psychological-projection/

39 "How American Racism Influenced Hitler," The New Yorker, accessed July 31, 2019, https://www.newyorker.com/magazine/2018/04/30/ how-american-racism-influenced-hitler.

40 "The American Indian And 'The Great Emancipator,'" United Native America, accessed August 22, 2019,. http://www.unitednativeamerica. com/issues/lincoln.html

41 "Dakota Access Pipeline Protest," Huffington Post, accessed July 30, 2019, https://www.huffpost.com/entry/dakota-access-pipeline-protest-photos.

SIX. INDIGO: FAITH

42 "Laudato Si'," The Pope's Climate Encyclical, accessed July 31, 2019, https://w2.vatican.va/content/francesco/en/encyclicals/documents/ papa-francesco.

43 "Faith Based Statements on Climate," Citizens Climate Lobby, accessed July 29, 2019, https://citizensclimatelobby.org.

44 "Hindu Declaration On Climate Change," Hinduismtoday.com, accessed July 29, 2019, https://www.hinduismtoday.com/pdf_ downloads/hindu-climate-change-declaration.pdf.

45 "The Time To Act Is Now A Buddhist Declaration On Climate," accessed July 29, 2019, http://fore.yale.edu/files/Buddhist_Climate_ Change_Statement_5-14-15.

46 "Jewish Groups Add Voices To Green Concerns,"*Reuters.com*,accessed July 29, 2019, http://blogs.reuters.com/environment/2008/08/08/jewish-groups-add-voices-to-gree

47 "Sources Of Our Living Tradition," UUA, accessed July 29,2019, https://www.uua.org/beliefs/what-we-believe/sources

48 "Pope Francis Tells Congress: Be Courageous, Do Something," *Newsweek*, accessed July 31, 2019, https://www.newsweek.com/pope-francis-climate-change-congress-376019.

49 "Threat Of Global Warming/climate Change | Social Justice," UUA, accessed July 31, 2019, https://www.uua.org/action/statements/threat-global-warmingclimate-change.

50 "Rise for Climate Organizing Guide," Rise For Climate.org, accessed July 29, 2019, https://riseforclimate.org/organising-guide/

SEVEN. VIOLET FUTURE

51 "Climate Change Skeptic Group Seeks to Influence 200,000 Teachers," *PBS*, accessed August 31, 2019, https://www.pbs.org/wgbh/frontline/article/climate-change-skeptic-group-seeks-to-influence-200000-teachers/

52 "The Children's Climate Case: Our Obligation To Future," *The Hill*, accessed July 31, 2019, https://thehill.com/opinion/energy-environment/415307-the-childrens-climate-case.

53 "Our Children's Trust: Kelsey Juliana," *Our Children's Trust*, accessed July 30, 2019, https://www.ourchildrenstrust.org/kelsey

54 "Landmark U.S. Federal Climate Lawsuit", Our Children's Trust, accessed August 22, 2019, https://www.ourchildrenstrust.org/juliana-v-us.

55 "Since Our Leaders Are Behaving Like Children," *Clean Technica*, assessed August 31. 2019, https://cleantechnica.com/2019/01/25/since-our-leaders-are-behaving-like-children-we-will-have-to-take-the-responsibility-they-should-have-taken-long-ago/

56 "Our House is on Fire," *The Guardian*, accessed August 31, 2019, https://www.theguardian.com/environment/2019/jan/25/our-house-is-on-fire-greta-thunberg16-urges-leaders-to-act-on-climate.

57 "Democrats Should Be Learning from Alexandria-Ocasio-Cortez," Medium.com, accessed August 30, 2019, https://medium.com/s/story/democrats-should-be-learning-from-alexandria-ocasio-cortezs-communications-style-not-dismissing-9b9bd0c24dc3.

BIBLIOGRAPHY

Carter, Peter and Elizabeth Woodworth. *Unprecedented Crime*. Atlanta: Clarity Press, 2018.

Community Environmental Legal Defense Fund. *On Community Civil Disobedience in the Name of Sustainability*. Oakland: PM Press Pamphlet Series No. 0013, 2015.

Flannery, Tim. *The Weather Makers*. New York: Grove Press, 2005.

Gardiner, Stephen. *A Perfect Moral Storm*. New York: Oxford University Press, 2011.

Gore, Al. *An Inconvenient Truth*. New York: Rodale, 2006.

Grinspoon, David. *Earth in Human Hands*. New York: Hachette Book Group, 2016.

Hartmann, Thom. *The Last Hours of Ancient Sunlight*. New York: Three Rivers Press, 1998.

Hawken, Paul. *Blessed Unrest*. New York: Penguin Books, 2007.

Klein, Naomi. *This Changes Everything*. New York: Simon & Shuster, 2014.

Kolbert, Elizabeth. *Field Notes from a Catastrophe*. New York: Bloomsbury, 2006.

Marshall, George. *Don't Even Think About It*. New York: Bloomsbury, 2014.

McKIbben, Bill. *The End of Nature*. New York: Anchor Books, 1989.

Moore, Kathleen Dean. *Great Tide Rising*. Berkeley: Counterpoint, 2016.

Orestes, Naomi & Erik M. Conway. *Merchants of Doubt*. New York: Bloomsbury, 2010.

Pope Francis. *Laudato Si*. Vatican City: 2015.

Trout, Stefanie Brook, and Taylor Brorby, Pam Houston. *Fracture*. North Liberty, Iowa: Ice Cube Press, 2016.

Washington, Haydn and John Cook. *Climate Change Denial*. Abington UK: Taylor and Francis, 2011.

Whitney, Lynn & Ellie Whitney. *Faith Based Statements on Climate Change*. CreateSpace, 2012.

Wilson, Edward O. *The Meaning of Human Existence*. London: Norton, 2014.

Wood, Mary Christina. *Nature's Trust*. New York: Cambridge University Press, 2014.

ABOUT THE AUTHOR

A lifelong student of nature and human nature, Gail Collins-Ranadive is the author of eight books of creative nonfiction, three with Homebound Publications. She also writes the environmental column for *The Wayfarer*, a literary journal produced by Homebound Publications. Her writings reflect a need to make sense of life experiences through multiple lenses that include being an oldest sister, psychiatric nurse, military wife, mother of two daughters, poet, volunteer educator, peace activist, interim minister, and grandmother of five. She and her partner Milt spend their summers in Denver and winters in Las Vegas.

HOMEBOUND PUBLICATIONS
OFFERINGS

———

To Lose the Madness by L.M. Browning
The Voices of Rivers by Matthew Dickerson
Hiking Naked by Iris Graville
A Fistful of Stars by Gail Collins-Ranadive
How Dams Fall by Will Falk
The Comet's Tail by Amy Nawrocki
Painted Oxen by Thomas Lloyd Qualls
A Letter to My Daughters by Theodore Richards
The Ashokan Way by Gail Straub
Woodland Manitou by Heidi Barr
The Kiss of the Sweet Scottish Rain by Robert McWilliams
Canoeing the Legendary Allagash by David K. Leff
Letters from the Other Side of Silence by Joseph Little
Sheltered in the Heart by Gunilla Norris